MATHEMATICS EAST AND WEST

The Ages of Mathematics

VOLUME TWO

MATHEMATICS EAST AND WEST

Charles F. Linn

Solon High School Library

DOUBLEDAY & COMPANY, INC.
GARDEN CITY, NEW YORK

LIBRARY OF CONGRESS CATALOGING IN PUBLICATION DATA
MAIN ENTRY UNDER TITLE:

THE AGES OF MATHEMATICS.

INCLUDES INDEXES.
CONTENTS: V. 1. MOFFATT, M. THE ORIGINS.—V. 2.
LINN, C. F. MATHEMATICS EAST AND WEST.—V. 3.
COOK, C. C. WESTERN MATHEMATICS COMES OF AGE.—V. 4.
COOK, P. D. THE MODERN ATE.
1. MATHEMATICS—HISTORY. [1. MATHEMATICS—
HISTORY] I. LINN, CHARLES F. II. MOFFATT,
MICHAEL, 1944–
QA21.A35   510'.9
ISBN 0-385-11216-5 TRADE
0-385-11217-3 PREBOUND
LIBRARY OF CONGRESS CATALOG CARD NUMBER 76-10336

COPYRIGHT © 1977 BY DOUBLEDAY & COMPANY, INC.
ALL RIGHTS RESERVED
PRINTED IN THE UNITED STATES OF AMERICA
FIRST EDITION

*For
My
Parents,
Helen and Charles Linn*

CONTENTS

1	*The Greek Heritage*	1
2	*The West's Asleep*	13
3	*Mathematics East*	26
4	*Farther East—India and China*	47
5	*The West's Awake*	65
6	*West Meets East*	81
7	*Numbers Are for Multiplying*	103
8	*From Renaissance to Renaissance*	122
	Epilogue	143
	Index	146

1

THE GREEK HERITAGE

Euclid, so the story goes, was leaving the Library at Alexandria, papyrus scrolls under his arm, when he was stopped by a street urchin.

The young fellow said, "What do you do in that building?"

"I am writing a book," replied Euclid.

"Read me some of your book," demanded the urchin. So Euclid unrolled one of the papyri and began to read what (traditionally) has been considered the opening statement in the geometry.

"A point is that which hath no part." Before he could continue the lad interrupted.

"Has a point got a smell?" he objected.

"No," the master was forced to admit.

"Has it got a taste?" And again Euclid had to demur.

"Then," demanded his tormentor, "why don't you put in your book that a point has no taste and no smell either?"

Euclid, finding himself frustrated by this unreasonable logic, shouted at the lad, "Begone. Mathematics is not for you."

And, mathematics was *not* for him, for, after all, the street urchin could hardly aspire to the kind of leisure that was required for serious contemplation of geometrical matters. Besides, he was much too literal-minded. In those days, it was all very well for the astronomer and musician to get involved in the practical implications of mathematics, but in geometry, the theory was the important thing.

Until very recently, when mathematicians became distressed with certain flaws in Euclid's system, the definition still stood—a point is that which hath no part. This satisfied the pure mathematician. And the practical man knew that a point was a dot on a piece of paper and he did not really worry about whether or not it had a part.

Whatever the shortcomings, and they were remarkably few, of Euclid's great compilation of mathematical results, his name was practically synonymous with "mathematics" through the nineteenth century. In particular, for the period of time covered in this book, his name heads a very short list that includes Pythagoras, Archimedes, Diophantus, and Ptolemy.

This arbitrary choice may stir up some argument. The historian of mathematics may contend, for instance, that Thales was certainly one of the most important in founding this Greek "tradition." Mathematicians will contend that Apollonius should not be slighted. If you were interested in the history of astronomy, you probably would argue that Hipparchus may very well have contributed as much as Ptolemy to the development of that branch of mathematics.

But this selection is based on the impact these mathematicians had on society in general, down to the present, and in particular in the period from about 400 to the beginning of the Italian Renaissance in the West. This is not a matter of mere familiarity, but rather that these five men had more effect than other mathematicians on the man in the street.

These men epitomize the Greek legacy of mathematics to the Moslems and later to the Western Europeans. They represent the several periods of intensive mathematical investigation: the time prior to the founding of Alexandria; the first Alex-

The Greek Heritage

andrian period, which lasted about a hundred years; and the second Alexandrian period, which followed the subjugation of Egypt by Rome, and the re-establishment of order. The influence of these mathematicians has been of mathematics rather than of personalities, for practically nothing is known of the men themselves.

There are a few facts available on the life of Archimedes. At least we can fix the date of his death in the Roman sack of Syracuse in 212 B.C. He was probably the most versatile of the Greek scholars—and perhaps of all recorded history. His accomplishments are generally hailed by mathematicians and physical scientists, as well as nonmathematicians. Voltaire, for example, after conceding that "There is an astonishing imagination even in the science of mathematics," went on to say "there was far more imagination in the head of Archimedes than in that of Homer." This is quite a concession from a literary figure who was not known for his fondness for mathematics.

Archimedes, in contrast to Euclid, was an originator of mathematical results and was less concerned with deductive logic than with an experimental approach to problems of pure and applied mathematics. This spirit of experimentation is in keeping with the tradition of the Pythagoreans, who applied it to all four branches of mathematics.

The figure of Pythagoras is almost legendary. Some historians even doubt that "Pythagoras" was one individual, and it is generally agreed that results credited to Pythagoras were the collective work of the school of which he was the founder. There can be no doubting, however, the impact of the Pythagoreans on the mathematics of the latter-day Romans, the Moslems, and the Western Europeans.

Euclid probably lived at the time of the founding of the Library and Museum at Alexandria. His name is linked, again in anecdotes and legends, with that of Ptolemy I (Soter), who was the first King of Egypt after the breakup of the empire of Alexander the Great. Euclid, so the story goes, assured his royal pupil that there was no short cut for him to a mastery of mathematics. A prince must apply himself as diligently as would any aspiring scholar.

The name of Euclid is, however, probably the best known of all mathematicians. His fame is due not to his originality as a mathematician, but rather to his *Elements*, in which he organized the mathematical results of his time. Euclid's geometry has been *the* geometry until our own time, and modern mathematicians use the word "Euclidean" in referring to a particular kind of geometry. The Arabs and the Jews translated the *Elements*, studied it, and made every effort to improve upon it, as did the later scholars of Western Europe. But Euclid's geometry stood the tests of many centuries.

About the year 500 the Roman grammarian Metrodorus collected forty-six algebra problems in epigrammatic form. Among these was the following:

> Diophantus passed one sixth of his life in childhood, one twelfth in youth, and one seventh more as a bachelor. Five years after his marriage was born a son who died four years before his father, at half his father's age.

If the last reference is to his father's age at death, the problem can be expressed as:

$$\frac{x}{6}+\frac{x}{12}+\frac{x}{7}+5+\frac{x}{2}+4=x,$$

from which you can figure that Diophantus was eighty-four when he died. This is all the information we have about the life of the greatest Greek arithmetician. Ask when he lived, and you can get almost any year between 150 B.C. and A.D. 350. In his *Arithmetica* he quotes Hypsicles. In turn Diophantus is quoted by Theon of Alexandria. The time of the former is in question, and the time of Theon is fixed only by his report of an eclipse in 364 . . . so you can see that the reasoning is very tenuous.

The Greek Heritage

There is a theory that "about 250" is the most likely time, but this is based on an admittedly questionable passage from an eleventh-century writing.

The two statements of the Diophantus epigram represent the first and third stages in the development of algebra—the rhetorical algebra and the symbolic algebra. The great contribution of Diophantus was in his development of the second stage, syncopated algebra, in which he abbreviated many of the more commonly used quantities and operations.

It is convenient to write "Ptolemy the astronomer" to avoid confusion with the many Egyptian kings of the same name, but Ptolemy, along with most of the other Greek mathematicians, studied seriously other areas, including music and geography. (His world map was considered a classic for at least a thousand years.) There is practically nothing known of his life. In his writing he dates himself to about A.D. 150 with recordings of eclipses that he observed. Using this same method, historians have been able to date others, Hipparchus and Menelaus for example, to whom Ptolemy acknowledges his debt. Even the title of his greatest work has been changed. Ptolemy called it *The Mathematical Collection*, but it is universally known as the *Almagest*, a Moslem contribution, and probably a distortion of a Greek word meaning "the greatest."

This change of title, and the general acceptance of the change, points up a major problem of the historian of mathematics. Many of the Greek works have come down to us only through translations. How much have the translators added or changed? How do you assign credit to an innovation? There are many examples of apparent contradictions and misnomers. The formula that most texts list as "Hero's" was evidently known by Archimedes some three hundred years before Hero. Even more dramatic are the cases of the "Pythagorean" theorem and "Pascal's" triangle.

Much of the work done by the Greeks in both geometry

and arithmetic reflected their interest in the mystical properties of numbers and geometric figures. This attitude is traceable to the Pythagoreans who attempted to explain the universe in terms of numbers.

If you accept the Pythagorean notion of a universe based on simple numerical relationships, you let yourself in for strange mystical and imaginative interpretations. The number seven, for example, rated a special place in the Pythagorean scheme because it was unrelated to the elementary geometric forms. In particular, seven was deemed to be ungenerated by members of the cosmic triangle, the right triangle having sides of three, four, and five units. "One"—the monad—was not a number, nor was two, but rather a link between one and the numbers. Three represented the first plane figure, the triangle, and four the first solid figure, the tetrahedron. Since $3^2 + 4^4 = 5^2$, the latter is accounted for, and six represents the area of the three, four, five right triangle. So it goes . . . but seven is ungenerated and, therefore, special. Incidentally, three is also referred to elsewhere as the "first number," including mention in Augustine's *City of God*.

In the realm of geometry the Pythagoreans made much of the regular solids—the tetrahedron, cube, octahedron, icosahedron and dodecahedron. Plato was likewise intrigued by these figures, which are frequently referred to now as the "Platonic solids." He recognized the first four and related them to the four elements, fire, water, earth, and air. Hard pressed to account for the dodecahedron, he finally decided to relate it to the all-encompassing heavens. Ridiculous? Hardly more so than the following statement:

> However, there are as it were two noteworthy weddings of these figures made from different classes: the males, the cube and the dodecahedron, among the primary; the females, the octahedron and the icosahedron, among the secondary, to which is added one as it were bachelor or hermaphrodite, the tetrahedron, because it is in-

scribed in itself, just as those female solids are inscribed in the males and are as it were subject to them, and have the signs of the feminine sex, opposite the masculine, namely, angles opposite planes. Moreover, just as the tetrahedron is the element, bowels, and as it were rib of the male cube, so the feminine octahedron is the element and part of the tetrahedron in another way; and thus the tetrahedron mediates in this marriage.

which comes from Johannes Kepler's *Harmonies of the World*. Kepler also devised a model of the planetary orbits based on the five regular solids, and ignored completely some very obvious discrepancies between his model and astronomical observations.

Of Pythagorean origin also, and featured in Euclid's *Elements*, is that which is usually called now the "golden section." This refers to the division of a line segment such that the ratio of the smaller part to the larger is equal to the ratio of the larger part to the whole segment. This ratio was much used by the Greeks in their architecture, and in their sculpture. Kepler's observation on the golden section was not quite so mundane as that last quote. "Geometry has two great treasures, one is the theorem of Pythagoras, the other the division of a line into extreme and mean ratio; the first we may compare to a measure of gold, the second we may name a precious jewel."

On the other hand there are enough demonstrable results related to the Pythagorean number philosophy that you cannot write it off as sheer nonsense. For example, members of that mystical brotherhood were able to verify experimentally the existence of simple ratios relating notes of the musical scale. Kepler's investigation of the properties of the regular solids and similar numerical ideas are said to have led him to the formulation of his three laws of motion, which are all important in the theory of planetary orbits. The three laws do show simple numerical properties in the movements of the planets. Psychologists have shown that observers are attracted to rectangles whose measurements involve the golden section—the "most pleasing rectangle."

Astronomy had its mystical side in astrology, based on the belief that the heavenly bodies, in particular the sun, moon, and planets, had an observable influence on the lives of men. In fact, a great deal of the investigation of astronomical phenomena by the Babylonians, Egyptians, Greeks, Moslems, and later Europeans of excellent reputation was inspired by the need for a more reliable basis upon which to predict these influences. In the Greek era and the "Middle Ages" astronomers were also astrologers. As such, they were regularly consulted by statesman and commoner alike. The astrological tradition continues, though it is presently in very poor repute, and certainly is now not acknowledged by astronomers, much less mathematicians.

Can you write off completely the notion that there is interaction between the activities of men and the movements of the planets? The alchemist was long ranked with the astrologer as the butt of ridicule, but he has been vindicated by modern atomic physics. Our outlook on witchcraft, ghosts, and the like has been modified by recent research in parapsychology. Perhaps the ancients (and latter-day adherents) were not completely wrong on the matter. Whatever your assessment, the fact remains that these notions were important to the Greeks and to their successors in the development of our mathematical tradition. Pythagorean mysticism prevailed along with the Pythagorean spirit of experimentation. Astronomy was mathematics, and astrology was, as Kepler observed, a foolish but necessary daughter of astronomy.

How different might be our outlook and how much richer this mathematical heritage had the original written sources or contemporary commentaries survived. That they did not is certainly not the fault of the mathematicians, and this brings us to a consideration of what might be described as politics and mathematics.

One of the most dramatic incidents in the history of the an-

The original title of this drawing is simply The Astrologer. *If it resembles other drawings of astronomers, cartographers, and mathematicians, this is not surprising in the least. For hundreds of years, the same people did all these things. But my impression is that astrology was the most profitable and provided the steadiest employment. The artist was Johannes Scherr.* Courtesy of the New York Public Library picture collection

cient world was the destruction of the Library at Alexandria. As in the much-discussed death of Archimedes, Roman soldiers were involved. What must have been Caesar's reaction when word reached him that the fire he ordered set had burned the Library? A fortune, or misfortune, of war? Julius Caesar was a soldier and he had ordered the burning of the Egyptian fleet to reduce the pressure on his Roman legions, who were in Egypt settling the civil war between brother and sister of the Egyptian royal family. Caesar could hardly be blamed for the spread of the fire. On the other hand, Caesar was an educated man whose writings were to become classics in their own field. He could not have been other than overwhelmed by the thought of the loss of hundreds of thousands of volumes methodically collected over a period of three centuries.

Would it be sheer rationalization for Caesar to note that with the increased chaos in Egypt, were the Romans to be defeated, the Library might very well have succumbed anyway? Indeed, there had been little intellectual activity in Alexandria during the preceding two centuries. The re-establishment of peace and order under the Romans could, and in fact did, provide a setting for a revival of Alexandrian culture.

The sister in the civil war was the Cleopatra of Shakespeare's *Antony and Cleopatra*. She proved to be an intelligent ruler, speaking the several languages of her subjects, and a sincere patron of the arts and learning. Among other accomplishments, she inveigled Mark Antony into bringing the books from the library at Pergamum to Alexandria, thus partially compensating for Caesar's fire. Cleopatra, for all her much publicized personal immorality, must be given credit for helping to initiate the second great Alexandrian period.

The first flourishing of Alexandria had followed the efforts of another general, Alexander's lieutenant, Ptolemy, in establishing conditions in which scholars could work without restraint or concern for matters other than their academic pur-

The Greek Heritage

suits. This era lasted for three generations of his family, which is, I guess, about as much as you can expect from one family.

The fourth Ptolemy was of a different mold. His chief interest was in pleasure, and the riotous living of the ruling class set an example for the more lowly born, and rapidly dispelled the atmosphere in which intellectual pursuits thrived. Many of the scholars fled the city. Rival cultural centers developed at Rhodes, Pergamum, and Rome. Disorder in Alexandria continued until the dramatic intervention of Caesar and his legions.

Religious conflict brought the second Alexandrian period to an end. A Christian colony had been established there early in the first century, and it began to flourish as the order and justice of the best Roman era deteriorated into civil wars, reigns of local tyrants, and increased burdens on the individual. The promise of a better life hereafter appealed to the oppressed masses, and their response to Christian instigators was not different from the actions of the present-day oppressed. The scholars and their school were natural targets of the ensuing violence, which culminated in the destruction of the temple of Serapis, which housed the Library. This was about 389. In 418 a mob, urged on by Cyril, Christian Bishop of Alexandria, murdered Hypatia, daughter of the commentator Theon and herself the head of the school.

Hypatia's life and brutal murder have been the basis for romantic fiction. We do know from her correspondence with a student, Synesius, who later became bishop, that she was "beautiful, learned, and eloquent." Her commentaries on Apollonius and Diophantus helped to preserve their work. With her death came the end of the Alexandrian era.

Professor T. Dantzig says that "Mathematics flourished as long as freedom of thought prevailed; it decayed when creative joy gave way to blind faith and fanatical frenzy." It is hardly as simple as that. We have seen in our own time that the oppressive impact of fanatical frenzy does not completely stifle the creative force. Consider, too, the time span which produced these thirty or so mathematician-astronomers of the Greek era—about a thousand years separated Pythagoras from Pappus. There was at least one lengthy period of almost three hundred years which was hardly more productive than were the two centuries following the death of Hypatia. And there were shorter periods of apparent intellectual stagnation.

The difference here is that a remarkably vital new force appeared on the stage of history and seized the initiative from the Greeks. This new religious-political force was Islam.

2

THE WEST'S ASLEEP

Frequently the historian will write off the A.D. 500–1000 era as the "Dark Ages," at least in Europe. He describes the rise of the monasteries, dwells at some length on the ebbings and flowings of the barbarian hordes, mentions the advent of Charlemagne, and then hurries on to the Great Renaissance. On the other hand, it seems reasonable to draw an analogy between Europe of that time and Africa of the present, and characterize early medieval Europe as "emerging." Whether you call it the "Dark Ages" or the "emerging" or the "period of fallowness," there can be little argument about the general lack of real intellectual and mathematical activity.

But here lies a good reason for looking at the period in terms of interest in mathematics. It is a chance to see how the "nonmathematician" regarded the business of numbers and quantitative measure—to see what, in the words of Queen Guinevere to King Arthur in the play *Camelot,* "the simple folk do." Not surprisingly, they looked at mathematics about as does the man in the street of our own time.

There are a few individuals who stand out in this period—Boethius the Roman, the English monk Bede (called "the Venerable"), Alcuin, another English monk, who became adviser and tutor to Charlemagne. The men themselves are obscure, for the principal accounts of the time were concentrated on religious activity. But enough can be gleaned from their own writings, from letters of others, including Charlemagne himself, and from collections of puzzle-problems, to give you an idea of mathematics in "emerging Europe."

Since the collapse of the Roman Empire and Roman order is said to have signaled the beginning of the "Dark Ages," it seems fitting that this account of the era, from A.D. 500 to 1000, should begin and end in Rome.

Anicius Manlius Severinus Boethius, Roman consul early in the sixth century, was described by Gibbon as "the last of the Romans whom Cato or Tully could have acknowledged as their countryman." He wrote on arithmetic—really what we now call "theory of numbers"—a geometry, and an important book on music. Music was at that time considered a part of mathematics. The Boethius work on music, incidentally, was long used as a text at Oxford and Cambridge, and still is a valuable reference on ancient music.

His arithmetic has often been called a translation of the work of Nichomachus (about A.D. 100), but it is much more original than that. He notes in the preface that he has taken certain liberties with the Greek arithmetic—abridging when he thought Nichomachus repetitious, and adding formulas and diagrams of his own when he thought they would clarify points in question.

Boethius wrote about such things as figurate numbers. Figurate numbers still fascinate many casual students of mathe-

---

Opposite, this illustration spans almost a millennium and a half. The subject is the philosopher-mathematician Boethius, sometimes called "the last of the noble Romans," whose writings on number theory and music were read at universities well into the seventeenth century. The essay for which the illustration was done was by Geoffrey Chaucer, the fourteenth-century poet and author of The Canterbury Tales. *The actual illustration is from the Kelmscott* Chaucer, *the crowning achievement of William Morris, who revolutionized the art of printing in the nineteenth century. Courtesy of Dover Publications*

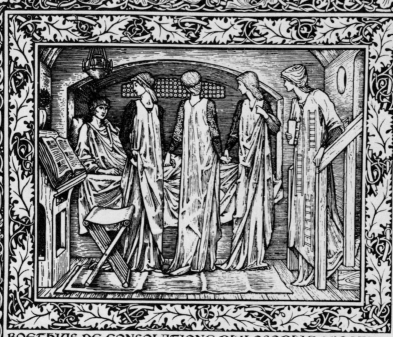

BOETHIUS DE CONSOLATIONE PHILOSOPHIE. BOOK I.

Metre I.
Carmina qui quondam studio florente peregi.

I, WEPING, AM CONSTREINED TO biginnen vers of sorowful matere, that whylom in florisching studie made delitable dittees. for lo! rendinge Muses of poetes endyten to me thinges to be writen; and drery vers of wrecchednesse weten my face with verray teres. At the leeste, no drede ne mighte overcomen tho Muses, that they ne weren felawes, and folweden my wey, that is to seyn, whan I was exyled; they that weren glorie of my youthe, whylom weleful and grene, comforten now the sorowful werdes of me, olde man. for elde is comen unwarly upon me, hasted by the harmes that I have, and sorow hath comaunded his age to be in me. Heres hore ben shad overtymeliche upon myn heved, &the slake skin trembleth upon myn empted body. Thilke deeth of men is weleful that ne cometh not in yeres that ben swete, but cometh to wrecches, often ycleped.

LLAS! allas! with how deef an ere deeth, cruel, torneth awey fro wrecches, and naiteth to closen wepinge eyen! Whyl fortune, unfeithful, favorede me with lighte goodes, the sorowful houre, that is to seyn, the deeth, hadde almost dreynt myn heved. But now, for fortune cloudy hath chaunged hir deceyvable chere to meward, myn unpitous lyf draweth along unagreable

matics. You have the "triangular numbers":

1,   3,   6,   10,   15,   21,   etc.,

and, of course, the "square numbers":

1,   4,   9,   16,   25,   36,   etc.,

and, in fact, a sequence associated with each regular polygon. Another topic discussed in some detail was the star polygon. Begin by marking off equal intervals around a circle, and then join the points in a systematic fashion. If you join adjacent points you get the regular polygons. If you alternate points you get the star polygons. These topics are straight from the Pythagoreans. The star polygon obtained by connecting every other point of a circle divided into five intervals is the familiar "star" of many a doodling, and was the symbol—the mystic pentagram—of the Pythagoreans.

Boethius was a friend and adviser of Theodoric, the Ostrogoth ruler of Rome, but eventually fell from favor, apparently because of the jealousy of Theodoric's subordinates. Imprisoned, he wrote his greatest work, *The Consolation of Philosophy*, which was completed shortly before his execution. This work shows that Boethius was not a Christian, though later writers described him as a martyr to the cause of orthodox Christianity. His remains were enshrined, some five hundred years later, beneath the high altar of the cathedral at Ciel d'Oro, with an inscription written, appropriately enough, by the mathematician Gerbert, who became Pope Sylvester II.

The reason for this strange turn of events is worth noting, since the religious theme is so dominant at this time. Theodoric was an Arian, as were most of the Ostrogoths. They had been converted by missionaries of the Arian sects, who rejected the divinity of Christ. Orthodox Christians, led by the emperor Justin, were feuding with the Arian Christians at the time, and

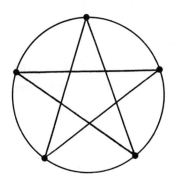 

Star Polygons

found it convenient to embrace one of the last of the noble pagans as a victim of their religious rivals.

Music theory, as advanced by Boethius, reflected the Pythagorean belief that all physical phenomena could be explained in terms of simple numerical relationships. The basic ratio was that of the fifth—the jump from C to G. This can be shown on the monochord, a one-stringed musical instrument used by the Pythagoreans, by placing the bridge so that it divides the string in a ratio of 2 : 1. That is, the ratio of the longer subdivision to the original length is 2 : 3. Dividing the string in a ratio of 3 : 4 produces the fourth—the C to F span. If the bridge is placed so that the original string is bisected, that is, the new length to be tested is half the original, the tone from the half is an octave above that of the original.

If, from C you move up the scale two fifths, you reach D. Drop down an octave then, and you have the note right next to the prime C. That is, two-thirds of the string, and then two-thirds of that gives you the high D. Double that length to reduce the tone by an octave . . . and you have that single "tone." The Pythagorean diatonic scale, as described by Boethius is built up in this way, to give the following ratios associated with a scale having C as the prime:

C	D	E	F	G	A	B	C
1	8/9	64/81	3/4	2/3	16/27	128/243	1/2

You could, of course, note that the ratio of the fifth to the fourth, $\frac{2}{3} : \frac{3}{4} = \frac{8}{9}$ . . . the ratio of the tone. Then insert two tones between C and F; two more between G and high C, to get the same ratios.

Now you may argue that these aren't exactly "the simplest possible ratios" claimed by the Pythagoreans. And indeed, modern theory, propounded most thoroughly by Helmholtz in the nineteenth century, holds the following simpler ratios:

1,   8/9,   4/5,   3/4,   2/3,   3/5,   8/15,   1/2,

which are not far removed. The theory, as transmitted by Boethius, was not really improved upon until Galileo's experiments in the sixteenth century. Boethius also subscribed to the Pythagorean notion of relating the scale ratios to the ordering of the planets, the sun, and the moon—a natural for them, since there were seven heavenly bodies and seven notes. Kepler, some 1,200 years later, attempted to establish such relationships.

The monasteries were almost completely responsible for the transmission of the culture of this time. The idea of these religious centers was not originally Christian. There were many enclaves of holy people in India long before the Christian era began, and such groups as the Essenes were recognized in Palestine. These early monastic groups stressed mysticism and asceticism.

St. Benedict adapted the idea to Christian organizations, but stressed communal effort and service, rather than extreme asceticism. Much effort was devoted to the copying of books, translations, and teaching—particularly to prepare novices in the monasteries. For several centuries the Benedictine monasteries, which operated independently, dominated the scene in Europe.

Another group of monastics were the Irish. Their origins are obscure, probably related to the tribal system, and more closely akin to the Oriental idea of the monastery. But they sent forth their missionaries, including St. Columban, who founded monasteries in England. A theory is advanced that the Irish priests even reached North America in the eighth century. The Benedictines, moving west, met the Irish monks venturing east in the seventh century, with the Benedictines emerging in the dominant role. By this time several monasteries were well established in England, and in one of these, at Jarrow, the monk Bede labored. From another of these monasteries, York, the monk Alcuin was called by Charlemagne.

Bede was educated at Jarrow and spent most of his life in the monastery, as he described it, "to learn, to teach, and to write." His greatest work was the *History of the Church of the English People*, which is remarkable just for the fact that he thought of the English as one people. This was at a time not too far removed from that of the legendary King Arthur, some 350 years before the arrival of the Normans, and when the island was still divided into a number of small antagonistic kingdoms.

Bede's writings on scientific matters, as he saw them, reflected the concerns and activities of the people of his time. These included observations of tidal waters, diagrams, and descriptions of the movements of the moon and directions for its use by "those unskilled in astronomy," and a thorough description of the system of digital reckoning.

There are many earlier references to digital reckoning and representation of numbers. Pliny the Elder, in his *Natural History*, tells of the dedication of a statue of two-faced Janus which had the fingers in a position to show 365, "and thus to represent him as the god of time and duration." Quintilian, after extolling the virtues of geometry, acknowledges the need to know something about numbers. "For pleading cases in court it is very often in request. On these occasions, to say nothing of becoming confused about sums, if a speaker, by any uncertain or awkward movements of the fingers, differs from the accepted mode of calculation, he is thought to be poorly trained."

These primitive-looking structures, which resemble beehives (or ovens), were said to be the abodes of Irish hermit-monks in the early days of Christianity in Ireland. Later, some of these Irish monks are thought to have reached the "New World" in the ninth century, and quite an advanced civilization developed in Ireland, though they seem to have done little in the way of mathematics and science.

The West's Asleep

Bede did not speculate on the origin of these techniques, but they apparently are linked to the Romans. Similar multiplication techniques were used in widely separated areas that were governed by Rome. By contrast, there is no trace of finger computation among the Hindus.

Bede's observations on the determination of Easter reflected a general concern of the early Christians, for many of the other festivals were set according to the date of Easter. The celebration was a carry-over from the Jewish Passover, and as such its date depended on positions of both moon and sun, in particular the full moon after the vernal equinox. Christians of Jewish descent debated with Christians of Gentile descent about whether the occasion should be set for a Sunday, until the matter was supposedly settled at the Council of Nicaea in 325. But Bede notes that, as late as the middle of the seventh century, the King and Queen of Northumbria, who acknowledged different churches, celebrated Easter on different dates.

Even after the Nicaean decree that Easter would be on a Sunday and must follow the fourteenth day of the moon and that the vernal equinox would henceforth be on March 21, there were difficulties. The time of the new moon of which the fourteenth day follows the vernal equinox varies according to longitude. And, in order to make long-range determination of Easters, the Christians looked to astronomers for cycles relating the positions of sun and moon. The Jewish cycle of eighty-four years was used, and then a more accurate cycle of 532 years was adopted. Finally a nineteen-year cycle was settled upon, though this requires application of the Golden Number and Sunday Letter, as explained in the Book of Common Prayer.

Bede attempted to bring order out of the confusion by specifying the way in which Easter was to be determined. He listed the dates for Easter through the eleventh century.

When Charlemagne became King of the Franks late in the eighth century, the general level of intellectual activity in most of continental Europe was very low indeed. This state of affairs cannot really be blamed entirely on the invasions of the barbarians, for, aside from a few outposts of the Roman Empire, for example, Cologne, Trier, Metz, there never had been much concern for such things as mathematics. Charlemagne, really a barbarian, though a Christian convert, and unschooled himself, was determined to do something about this. So, to his capital at Aix-la-Chapelle he summoned such men as Paul the Deacon from Monte Cassino; Peter of Pisa; Dungal, the Irish monk, who was charged with reformation of the calendar; and Alcuin of York.

The last mentioned was the leading contributor to what is called the "Carolingian renaissance." He composed textbooks on the seven liberal arts, which included, of course, arithmetic, geometry, and music. He was the author of a collection of letters which is a primary historical source of the time. And, apparently, he made a collection of mathematical problems, for he refers to them in his letters. Though Alcuin's authorship of *Problems for the Quickening of the Mind* is disputed, there is general agreement that the problems come from that period of time and as such afford some insight into the way Charlemagne and his contemporaries thought about mathematics. Some of these puzzles are of the types found in present-day puzzle collections.

1. A dog chasing a rabbit that has a start of 150 feet, jumps 9 feet every time the rabbit jumps 7. In how many leaps does the dog overtake the rabbit?
2. A wolf, a goat, and a cabbage must be moved across a river in a boat holding only one besides the ferryman. How must he carry them across so that the goat shall not eat the cabbage, nor the wolf the goat?
3. If 100 bushels of corn be distributed among 100 people in such a manner that each man receives 3 bushels, each

woman 2, and each child ½, how many men, women, and children were there?

And here is one that may very well have come from the Moslems; it is much like the inheritance problems in Al-Khwarizmi's *Algebra:* A dying man wills that if his wife, being with child, gives birth to a son, the son shall inherit ¾ and the widow ¼ of the property; but if a daughter is born she shall inherit $7/12$ and the widow $5/12$ of the property. The poor lady bore twins, a boy and a girl. How then should the property be divided?

Activities of individuals during the early Middle Ages are not well documented, and the ladies, in particular, were hardly mentioned. But there is evidence, though not ample, that there were active correspondents among the women. At least one, a nun named Hrotsvitha, wrote many plays that were well thought of at the time, and that now are considered to represent an important transitional period in the history of drama.

Hrotsvitha's efforts form a part of a brief literary renaissance under Otto the Great, who sought to restore the "glory that was Charlemagne." She is accorded the honor, along with numerous other playwrights, of having it said that Shakespeare "borrowed" from her works.

In these plays, Hrotsvitha shows an interest in the theory of numbers. For example, in one play the emperor Hadrian asks of Wisdom the ages of her three daughters, Faith, Hope, and Charity. Wisdom replies that the age of Charity is a defective even number, that of hope a defective odd number, and that of Faith an abundant number. To which Hadrian says, "What a difficult and tangled question has been raised about the mere ages of these girls."

This requires a bit of explanation. The Pythagoreans and their latter-day adherents, particularly Boethius, categorized a number according to its relationship to the sum of its factors.

For example, six is said to be a "perfect" number, for it is equal to the sum of its proper factors, that is,

$$6 = 1 + 2 + 3$$

A number is said to be "abundant" if the sum of its factors is larger than the number. A "deficient" number is larger than the sum of its factors. Thus 12, 18, 20, 24 are abundant numbers. But 14, 15, 16, and, of course, the primes are deficient. Judging from the line Hrotsvitha has Hadrian speak, this kind of information was not widely held. Hrotsvitha, incidentally, spoke of three perfect numbers other than 6: 28, 496, and 8,128.

The perfect numbers were general favorites among the mathematically inclined of the time. Alcuin was questioned by a student on a passage in the *Song of Songs* (ascribed to King Solomon), "Why are there only threescore queens, but four score concubines?" Alcuin replied, "Thus, even as the number six, with its parts in the order of the unit, is perfect, so also must be the number sixty, with its parts in the order of the ten."

There is no record of the student's reaction to this explanation, but note that Alcuin apparently thought in terms of a positional decimal numeration system.

This account, as predicted, ends in Rome, with mention of Gerbert, the mathematician-pope, who represents an awakening in the West, and his story will come later.

Opposite, Hrotsvitha, a learned nun of the Benedictine abby in Gandersheim, Saxony, presents a copy of her "hymn of praise" to Emperor Otto I, in whose honor it was written. Hrotsvitha (who lived in the tenth century) included mathematical illusions in a play, Sapientia, *and also wrote the perfect numbers, 6, 28, 496, and 8,128. This woodcut is by Dürer, who made significant mathematical contributions himself.* Courtesy Dover Publications

3

MATHEMATICS EAST

The era from about A.D. 500 to about 1100 was dominated culturally, intellectually, and commercially by "eastern" peoples. Mathematical developments of the time were almost entirely the results of their efforts. There were the Moslems, the Hindus of India, and the Chinese. (Japanese mathematics really dates back only to the seventeenth century.) Their efforts were going on at the same time, so that these chapters are not chronological, but rather geographic, and somewhat arbitrarily divided at that.

That Moslem civilization, which encompassed the Mediterranean world from Spain to Palestine and extended even to India and parts of what is now the Soviet Union, filled a vacuum created by the decline of the Greek center at Alexandria and the Roman Empire. It was a civilization of many paradoxes, which are important in a study of the history of mathematics.

The Moslems were non-Christian, but for hundreds of years Christians and Jews did very well under Moslem rule and in fact contributed significantly to the intellectual output of their rulers.

The Moslem world was created and dominated at first by Arabs, of whom one scholar says: "Honor and revenge were the keynotes of the pagan Arab's ideal; to be free, brave, generous, to return good for good and evil for evil with liberal measure; to hold equally dear, wine, women and war; to love life and not fear death; to be independent, self-reliant, boastful, and predatory; above all, to stand by one's kinsmen, right or wrong, and to hold the blood-tie above all other obligations." Yet a characteristic of the Moslem civilization was a well-defined bureaucratic class in whose interest much of the intellectual work was done.

The Moslems have been accused of destroying the centers of Greek culture, including the Library at Alexandria. On the other hand, much of the effort of their scholars was devoted to the translation and preservation of the words of the Greeks. It was largely through these translations that Greek mathematics was rediscovered by the West.

To appreciate, and possibly resolve, these contradictions, we must first consider the origin of the Moslem civilization.

The emergence of Islam, as the culture of the Moslems is known, literally happened "all of a sudden." Moslems mark 622 of the Western calendar as the beginning of their calendar. It was in this year that an obscure Arab fled from Mecca to Medina, taking with him a few followers who accepted his claim of divine inspiration. This man became known as Muhammed, the Prophet (also Mohammed or Mahomet), and his writings became the Quran (more often written "Koran") —the all-encompassing law of the Moslems. Written in Arabic, these words of Allah-God prescribed the behavior of the faithful to the finest detail.

Upon Muhammed's death, about ten years after the flight to Medina, his followers numbered only a handful. The words and promises of the scriptures were not attractive enough to draw the Arabs away from their tribal quarrels and tribal gods. Muhammed's successors proposed a plan of wars against non-Arab peoples. The prospect was appealing and by 642 the Arabs had conquered the Persians on one side and had captured

From this picture of a Moslem mosque you can get an idea of the remarkable geometric patterns and designs that they used for decoration, in place of the proscribed images of "any living thing."

Mathematics East

Alexandria on the other. The date, you note, is some two hundred years after the murder of Hypatia, and even farther removed from the destruction of the temple that housed much of the Alexandrian Library.

The Moslem Empire eventually extended to the Pyrenees, where expansion was halted, at least in part, by the forces of Charlemagne. (Some historians argue that the Frankish empire thrived only because it was the sole effective counterweight to the forces of Islam.) In the east, the mathematician Al-Biruni served a Moslem court in Afghanistan. The religion, as revealed in the Quran, dominated all phases of the life of the converts, who in turn controlled the empire. They literally lived their religion during the waking hours of the day.

Many Western writers, particularly those associated with the Christian church, have tended to play down the accomplishments of the Moslems. The basis of their prejudice may be the Moslem rejection of the divinity of Jesus and toleration of such things as slavery and polygamy, although the Moslems did not introduce these institutions and probably were being only realistic in tolerating them. It is unfortunate that such prejudices have carried over to assessments of scientific endeavors.

Moslem society, as it began to stabilize after the period of rapid expansion, was certainly cosmopolitan. Persians were quite prominent, bringing with them a tradition of efficient bureaucratic government, poetry, and literature, and the Zoroastrian religion. Omar Khayyam, poet and mathematician, was a Persian, as was Al-Biruni. Jewish scholars and physicians flourished, particularly in Spain. In the Near East, the oppression of the Byzantine church drove the Christians of Syria and the Coptic sect of Egypt to a rapprochement with the Moslems.

Most of these people retained their own religion and were socially of lower status than were those who became Moslems. Nevertheless, they were tolerated, at least until the twelfth century, and provided the Moslems with the intellectual tools they lacked.

Professor George Sarton notes of these Copts and Syrians, "It is not too much to say that without the unstinted help given by the oriental Christians the new Islamic culture would not have arisen. The best proof of their eagerness to help is that they lost, if not their souls, at least what was almost as precious to them, their own language, in the process."

The matter of language is particularly important to a consideration of the Moslem intellectual efforts. The Quran was written in Arabic. Thus, Arabic was the only acceptable language for the Faithful. Abraham Ibn Ezra (the "Rabbi ben Ezra" of Browning's poem) argues that the translation of Greek works into Arabic was done by non-Moslems, particularly Jews, because it was feared that if a Moslem should work with another language he might die and be denied the benefits promised by the Quran.

Ibn Ezra (twelfth century) traces the approval of the translations to a dream of the Moslem ruler Abbassid al-Saffah (late eighth century) in which he was instructed to entrust the work to a Jewish scholar. (A similar story is told of Nishirvan the Just, who ruled Persia in the sixth century.) Anyway, the translation was completed; the king was delighted and rewarded the Jew handsomely.

The Moslems were inclined toward belief in dreams. Another account, by Ibn al-Nadim, tells of a dream of the Caliph Al-Ma'mun (ninth century) which supposedly inspired the Caliph to sponsor an investigation and translation of Greek scientific works. "He saw a man white in complexion but with a reddish tinge, a high forehead, close-together eyebrows, bald, with grey eyes and pleasant features, sitting on his throne. Al-Ma'mun said, 'I found myself standing before him, filled with awe at his presence. I said, "Who are you?" He said, "I am Aristotle." I was delighted to hear this and said, "O wise man, may

Mathematics East

I ask you a question?" He said "Carry on!" I said, "What is beauty?" He said, "That which appears beautiful to the intellect." I said, "And then?" He said, "That which is beautiful according to the prescriptions of revealed religion." I said, "And then?" He said, "That which is beautiful in the eyes of the people in general." I said, "And then?" He said, "That is all." ' "

Apparently the dream ended here, but Ibn al-Nadim goes on to tell how the inspired Al-Ma'mun sent wise men to the Emperor of Byzantium to bring back knowledge of the ancient sciences.

As the Moslem society crystallized, portions of what is loosely described as "an empire" were ruled by independent caliphs. In each court there developed a class of "secretaries" who administered the government, collected taxes, planned and executed public works. These "secretaries" required the best possible preparation. It is largely in response to the needs and interests of this group that a special literary form, *adab*, was developed, a practical mathematics emerged, translations of the Greek works were encouraged, and Hindu and Chinese innovations reported.

A ninth-century writer, Ibn Qutaiba, describes some of the skills expected of the "perfect secretary": "The Persians always used to say, 'He who is not knowledgeable about diverting water into channels, digging out courses for irrigation streams and blocking up disused well-shafts; about the changes in the length of the days as they increase and decrease, the revolution of the sun, the rising-places of the stars and the state of the new moon as it begins to wax, and its subsequent phases; about the weights in use; about the measurement of triangles . . . such a person must be considered only partly qualified as a secretary.'"

The book *Keys of the Sciences* was written about a century later by Muhammed ben Ahmad Al-Khwarizmi to sum up the background required of the would-be secretary. This was a monumental work which included ordinary and special terms used by various specialists, discussions of scientific problems and basic principles of the various skills. (Incidentally, the author of *Keys of the Sciences* should not be confused with the ninth-century mathematician Muhammed ben Musa Al-Khwarizmi, although there is obvious justification for such an error.)

The Arabs were much concerned with classification of knowledge. They believed that the knowledge available to men was of a finite order "which a diligent scholar should, with God's guidance, be able to encompass."

Tenth-century Spain, under Hakam II, is an outstanding example of the cultural achievements of the Moslems. The university at Cordova, where Hakam had his capital, attracted thousands of students. The king was an enthusiastic bibliophile whose library is reported to have contained 400,000 volumes. He was supposed to have read all of them and to have annotated many in his own hand.

Hakam paid scholars just to come and study at his court. He established twenty-seven free schools in Cordova and paid the teachers out of his own pocket. It is said that in Spain "almost everyone" could read and write—a claim that must be suspect—but that such a claim could be seriously advanced and considered suggests a dramatic contrast with the rest of Western Europe.

As late as the fifteenth century, long after the decline of the Moslems as a political power, a devoted follower of the Prophet established a great cultural center in Samarkand. This ruler was Uleigh Beg, a theologian (he could recite, from memory, the Quran in all seven readings), patron of poets, historian (his principal work, *History of the Four Sons of the House of Ginqiz* [Genghis Kahn] has been lost), mathematician, and astronomer. He is best known for his work in astronomy. His book of astronomical tables was well known in Europe in the seventeenth and eighteenth centuries, and the observatory that he designed, regarded as one of the wonders of the world in his time, was only one of many magnificent buildings of Samarkand. Uleigh Beg, incidentally, ruled by the grace of the Mongols who dominated the country. His city was a meeting ground for the East and Far East.

It was in settings such as the courts of Uleigh Beg and Hakam that the mathematicians worked. Some of their best efforts were in response to the requirements of the astronomers of the courts, who were anxious to have the best possible information upon which to base their astrological predictions. Their writings reflect the dominant religious theme, particularly the words of the Quran.

Greek mathematics developed principally as a pure science. That is, the mathematicians were little concerned about the usefulness of their results. In the long run, Greek mathematical results, for example conic sections and indeterminate equations, proved to be very useful, and it is partly on this record of utility that the high regard for Greek mathematics is based.

Moslem mathematics, on the other hand, developed substantially in response to a variety of needs by the society. As the requirements changed, the utility of the results declined, and Moslem mathematicians have poor historical reputations. These needs ranged from the problems of weights, bridges, aqueducts, and accounting which Ibn Qutaiba mentions, through matters having a religious basis, to the requirements of the astrologers. Moslem mathematical efforts represent not a reaction against the Greek influence, as has been suggested at times, but rather a development in a totally different social context.

The devout Moslem was required, five times a day, to bow and pray in the direction of Mecca. Determination of this direction, or *kibla*, was considered quite important, certainly not to be left to the discretion of the individual. Much mathematical effort was devoted to its calculation for various locales, particularly such important cities as Bagdad and Cairo. This work yielded significant improvements in the methods of spherical trigonometry.

Concern for the kibla was not a Moslem innovation, incidentally. The early Hebrew kibla is mentioned in I Kings 8, 44, with Jerusalem the center of attention. The original Moslem kibla was also oriented to Jerusalem but later was changed, probably reflecting the Prophet's disappointment over his slight success with the Jews.

Computation of the kibla and problems of administering a far-flung "empire" naturally led to a concern for better maps. Moslem efforts produced real mathematical cartography that rivaled that of their contemporaries in China. These maps are a real contrast to those produced in Western Europe at the same time, and even up to the "Italian" Renaissance.

Dating from the period of decline of Moslem power there is the famous world map of Al-Idrisi, made about 1150 for Roger II of Sicily. Norman King Roger instigated the serious work of translating Moslem works into Latin, and in his time Sicily became an important center of East-West cultural transmission. This world map showed the influence of Greek cartography, as epitomized by Ptolemy's work, and of Chinese cartography. The latter is really not so surprising in view of the many Moslem-China contacts, including an Arab colony established in Canton in the middle of the eighth century.

Certain prescriptions of the Quran required mathematical treatment. Much of Al-Khwarizmi's algebra was devoted to resolving inheritance problems raised by the holy book. Careful astronomical calculations were needed for religious purposes. Ramadan, a period of fasting, began and ended with certain positions of the moon, for example.

The astrologers were in high repute, and most astronomers doubled as astrologers. Al-Biruni noted that this was essential, since the practicality of astrology was readily agreed upon, but a man would be suspect if he studied the heavens just for the sake

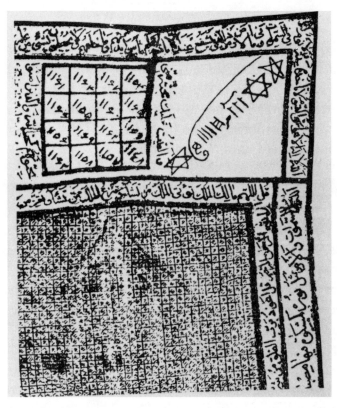

A portion of an Islamic magical manuscript, included in the David Eugene Smith Collection at Columbia University. You can see part of a very large magic square and, at the top left, an order four magic square. The diagonal line in the upper right is apparently a Moslem elaboration of the Hebrew "Seal of Solomon." Courtesy Scripta Mathematica

of studying the heavens. Whatever the rationale, the cause of mathematics benefited.

Serious attention was given to the translating of Greek mathematical works at Bagdad in the late eighth and early ninth centuries. The work, sponsored by the Caliph Al-Ma'mun, must certainly have been significant in its time, but the translators are obscured historically by the figure of Muhammed ben Musa al-Khwarizmi. Al-Khwarizmi, who is credited with establishing the name "algebra," made no mention in his writings of these translator contemporaries. His geometric works bear no resemblance to that of Euclid, of which there was a translation, nor was his algebra influenced by that of Diophantus, possibly because he knew no Greek. (Diophantus was not translated until the tenth century.)

Al-Khwarizmi was the first to teach algebra in an elementary form and for the sake of practical applications. In his *Algebra* he made no attempt to justify procedures or encourage reasoning. The rules were proclaimed as if they were divine revelations or traditions of the ancestors, which the student followed as a faithful believer. The proclamation of such a mathematical "law" was frequently followed by such advice as "Proceed according to this rule and you will succeed with the help of God."

This tradition also is apparent in Hebrew and Hindu texts and certainly was carried over to algebra texts of the twentieth century. But blind obedience to rules and rote memorization notwithstanding, Al-Khwarizmi's *Algebra* has a substantial impact on later European algebraic developments, through Latin translations by Gerard of Cremona and Robert of Chester in the twelfth century and through the writing of Fibonacci.

Al-Khwarizmi's lack of explanation brought harsh criticism of his "algebra of inheritance" by later commentators—but not because he didn't explain. They argued that he was wrong. Moritz Cantor, for example, contends "Thus the great mathematician, al Khwarizmi, in his cunning craft, succeeded in thwarting the law, defeating algebra and logic, conciliating the antagonistic schools of law, and assuaging the dumbfounded mathematicians of his time."

But Al-Khwarizmi's inheritance problems, and the matter of inheritance was most important to the Moslems, only reflected the rules established by the Quran. These were complicated, confusing, and often contradictory. Consider, for example, a woman who dies leaving her two daughters, her mother, and her husband. The law prescribes that if only one daughter survives and no son, she inherits ½ of the estate; if there are two or more daughters (and no sons) they receive ⅔; the husband gets ¼, and the mother ⅙. But ⅔ + ¼ + ⅙ = ¹³⁄₁₂—something of a problem. There is an amendment to the law, however, which says that the denominator in such a case is to be made equal to the numerator. Hence the husband gets ³⁄₁₃, and so on, an apparent contradiction.

Al-Khwarizmi obviously assumed that his readers would be familiar with the legal rules of inheritance, never dreaming that an infidel would read his "algebra of inheritance," much less read it without first familiarizing himself with what the Quran had to say on the matter.

The "Arab horse problem," which turns up these days in puzzle collections, is probably a descendant of Al-Khwarizmi's inheritance problems. The problem is posed something like this: An Arab died, leaving his seventeen horses to be divided among his three sons as follows: half to the eldest; one-third to the second, and one-ninth to the youngest. How did the sons resolve the obvious problems of carrying out their father's will?

The solution is also in the tradition of Al-Khwarizmi, and usually is given in the words of a wise old friend of the family (though no Quran reference is cited). The old friend added one of his horses to the seventeen, making the division easy—nine, six, and two. This left one horse, hopefully the friend's, and the arrangement satisfied all concerned, including, presumably, the legalists.

If Al-Khwarizmi was not influenced by the Greeks, who were his mathematical antecedents? Philological considerations point to the Assyrians. For example, the Arabic *"al-jabr,"* Latinized as *"algebra,"* is closely related to the Assyrian *"gabru"* or *"jabru,"* which probably meant science of equations. Al-Khwarizmi's method for solving equations was similar to the approach favored by the Babylonians.

Al-Khwarizmi's geometric techniques indicate that he was familiar with a Hebrew geometry called the *Mishnat ha-Middot*, dating probably to about 150 of the Christian era but not discovered until 1862. Or it may be that Al-Khwarizmi and the *Mishnat ha-Middot*, a "mishna on mensuration," favor a common source.

In any event, the writer or compiler of the book faced the same kind of problem as did Al-Khwarizmi—that of reconciling the word of the scripture with empirical data. For example, the Bible (I Kings 7,23) described a "molten sea" with diameter 10 cubits and circumference 30 cubits. The early Hebrew geometer noted this and suggested that 3 be considered the "practical" value of the circumference/diameter ratio, while $3\frac{1}{7}$ or $\sqrt{10}$ is the "exact value."

The Moslems and Jews, incidentally, did not have a complete monopoly on the mathematical work in the Near East. Bishop Severus Sebokht, a Syrian living in a monastery at Nisibis, on the Euphrates, in the eighth century, seems to have helped transmit the Indian ideas on calculation and their "nine symbols" to the Moslems. Nisibis was a center of East-West trade, which could account for Severus' familiarity with the mathematical ways of the Indians. He was, in turn, the teacher of such men as Athanasius of Balad, patriarch of the Jacobites, and George, "Bishop of the Arab Tribes," who became well known as translators and polygraphers—side lines of their ecclesiastical activity among the Moslems.

Severus wrote on the astrolabe and eclipses (ridiculing the popular dragon-swallowing-the-sun theory), and also got in his licks at the Greek scholars, whose traditional arrogance toward Syrian scholarship distressed him. He went so far as to claim invention of astronomy by the Syrians. The Greeks, he noted,

Mathematics East

were mere pupils of the Chaldeans of Babylon, and the Chaldeans were but Syrians.

Al-Biruni, a tenth/eleventh-century Persian, wrote in Arabic. He was the protégé of several princes and served them in a variety of ways apart from his mathematical works. Al-Biruni was a practicing astrologer, but also was entrusted by one potentate, Khwarizmshah, with delicate political missions because of his "golden and silver tongue." His political efforts evidently came to naught, for the ruler was assassinated, and his entourage, including Al-Biruni, carried off to another court in Afghanistan.

There he accompanied the Sultan Mahmud on military expeditions into India. He taught the Greek sciences and learned Indian and Sanskrit mathematics and literature, telling of them in his *Description of India*. The book became an important link between the mathematicians of India and China and those of the Arabs to the West.

Al-Biruni's *Chronology of Ancient Nations* was more of a mathematical work, giving detailed and technical descriptions of the time computing systems used by the Persians, Sogdians, Chorasmians, Jews, Syrians, Harramians, Arabs, Greeks, and Romans. The *Chronology* also contained a chapter on map projections, the "method of the astrolabe" (stereographic polar projection), central projection, and cylindrical projection (orthographic).

The author remarks on the complexity of the calculations involved in going from, say, Greek to Arabic chronologies or from Arabic to Persian, and adds, "In such cases the cleverness of the student will manage to solve the problem, although the calculations necessary for such a derivation may be very long," and ends with, "God helps to find the truth!" Could Al-Biruni have had his tongue in cheek?

In a curious digression from an account of chronologies, he discusses a chessboard problem that appears in many forms in different cultures. One variation has it that a king offered to reward the inventor of chess by giving him an amount of wheat to be determined as follows: one grain on the first square of the board, two on the second, four on the third, and so on, with the amount doubling each time. Al-Biruni gives the total amount as 18,446,744,073,709,551,615 grains, and in sexagesimal notation, 30.30.27.9.5.3.50.40.31.0.15.

Al-Biruni, incidentally, was not the only sure-footed Moslem mathematician who managed to survive the political downfall of his patron. Nasiruddim Tusi, who lived in the early thirteenth century, made the rounds from court to court. The Moslem governor of the province in which he lived had him kidnaped and delivered to the Grand Master of the Assassins—a fierce group of land pirates who were the chief power in northern Persia at the time. (The Grand Master of the Assassins wanted his court adorned with scholars, which was then the vogue in Oriental courts.)

Tusi persuaded the Grand Master to surrender to the Mongols—who promptly put him to death. But Tusi nimbly entered the service of the Mongol, Hulagu, by whom he was highly esteemed as an astrologer. Hulagu took the Moslem scholar along when he sacked Bagdad. Tusi assured his latest patron that no heavenly vengeance would befall him if he executed the Caliph of Bagdad, which he did.

Having thus resolved the political problems, Tusi settled down to some serious mathematics and astronomy, operating in an elaborate observatory that the Mongols sponsored. He wrote some important criticisms of the *Almagest*, and a treatise in which trigonometry was treated as a mathematical system in its own right, something that wasn't done in the West until the work of Regiomontanus at Nürnberg, some two hundred years later.

Omar Khayyam was probably the outstanding mathematician of the time, but better known through his poetry, as

The entrance to a Moslem mosque which emphasizes the symmetry and intricate geometric configurations that were a feature of Moslem art and architecture.

translated by Edward FitzGerald. This Moslem poet-astronomer-mathematician, who lived at the time of the First Crusade, wrote *The Rubaiyat*, which often tends to make light of the writer's mathematical accomplishments. For example, the lines:

> Ah, but my Computations, People say,
> Have squared the Year to human compass, eh?
> If so, by striking from the Calendar
> Unborn To-morrow, and dead Yesterday

give little hint of the fact that Omar devised a calendar more accurate than the Gregorian.

From another quatrain with mathematical overtones,

> For "Is" and "Is-not" though with Rule and Line
> And "Up-and-Down" by logic I define,
>   Of all that one should care to fathom, I
> Was never deep in anything but . . . Wine,

you would hardly guess that Omar examined and wrote on two problems that are frequently cited now as epitomizing modern mathematics. These are non-Euclidean geometry and the systematizing of the real numbers.

Commentaries on the Difficulties in the Postulates of Euclid's Elements point up a question which led to the formulation of the "non-Euclidean" geometries. Omar noted, as indeed did Euclid himself, that the fifth postulate: "Through a point not on a line, there is exactly one line which does not intersect the given line," was "weaker" than the others. He suggested two alternatives, of the order of:

(i) ". . . there is no line which does not intersect the given line";
(ii) ". . . there are at least two lines which do not intersect the given line."

In noting these alternatives, Omar anticipated the investi-

Mathematics East

gation of the eighteenth- and nineteenth-century mathematicians Saccheri, Gauss, Bolyai, Lobachevski, and Riemann.

His discussion of ratio as related to the limiting process anticipated the work of Stevin (about 1585), which, in turn, led to the nineteenth-century work of Dedekind and Cantor. The latter put the study of the real numbers on a very rigorous basis.

There is a tendency among "Omarphiles" to read all kinds of remarkable revelations into the quatrains. It has been suggested, for example, that the lines:

> We are no other than a moving row
> Of visionary Shapes which come and go
>   Round with this Sun-illumined Lantern held
> In Midnight by the Master of the Show.

indicate Omar's belief that the earth moved about the sun. This seems rather meager evidence, and there is nothing in his other known writings to substantiate the conjecture. This was a time when some Moslem and Jewish philosophers were rejecting the Ptolemaic hypothesis, but in favor of the even less satisfactory proposal of Aristotle.

On the other hand, Omar showed remarkable prescience of other modern notions, and he was less beholden to religious and philosophical constraints than were his contemporaries in other parts of the Moslem world.

These considerations of religious orthodoxy were becoming increasingly significant at this time, about 500 A.H. ("After Hegira," the Prophet's strategic withdrawal from Mecca to Medina), and contributed heavily to the decline of Islam.

Al-Ash'ari spelled out religious orthodoxy in the tenth century, though it was not until about two centuries later that the religious teacher Al-Ghazzali led an effective wave of reaction against the scientists and non-Moslems. The atmosphere of tolerance, in which the poet Dakiki could boast:

> Of all that is good and bad in the world,
> Dakiki has chosen four things to himself:
> A woman's lips as red as rubies, the melody of the lute,
> The blood-colored wine; and the religion of Zoroaster,

was changed to one in which the Jewish physician, Maimonides, felt it necessary to write a lengthy apology for his advice to the sultan at Cairo that he should indulge in the forbidden wine and music as a cure for his melancholy.

The twelfth century was a time of the increasing corruption of the bureaucracy and the general decline of order, which permitted conquest of much of the Moslem world by the Mongols and Tartars. But by this time, the West was beginning to awaken from the Dark Ages, and in Sicily and Spain scholars were beginning the work of translating the works of the Moslems into Latin as well as their translations of the Greek works.

The Moslems, then, must be credited, at least, with transmitting the Greek mathematics to the West. Their harshest detractors, and there are many, cannot say less. On the other hand, we can hardly say more. In mathematics and related areas there were few men of real genius. (Omar Khayyam seems to be a notable exception to the general rule.) This is not really surprising when you consider that the Arabs started from "ground zero." They were a group of seminomadic tribes, surviving in a harsh environment. They had less than three centuries from the stabilization of the empire to the hardening of religious

---

As I have noted on several occasions, Moslem art was confined, by scriptural constraint, to ornament and mosaic and the like, which involve no graven images. This illustration, opposite, shows one pattern that has shown up in floors, ceilings, bronze doors, and other places. Such patterns, now referred to as "tesellations," have become popular facets of contemporary "geometry for little kids," particularly in England.

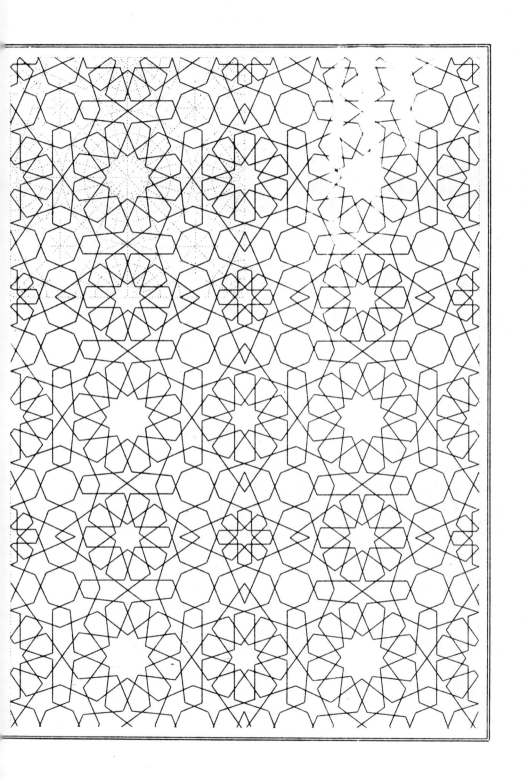

orthodoxy. Compare this time span to that in which the Greeks developed their mathematics.

It can be argued that Greek mathematics was available to the Moslems. But how available was it? There was hardly a tradition of Greek mathematics, for the Moslems and the Greek mathematics were far removed in time. Almost four hundred years elapsed between the decline of stability and productive work in Alexandria and the capture of the city by the sons of the Prophet. There was a language barrier, always a problem in the transmission of ideas from one culture to another, but accentuated here because of the religious significance associated with the Arabic. Then, too, the Greeks had evolved a leisure-oriented culture in which the contemplation of "pure mathematics" was a respectable preoccupation. The Moslems, with a tradition of combatting the environment, never quite saw it that way. Even after the caliphs began to surround themselves with scholars, nonpractical considerations were looked upon with suspicion.

In any event, the Moslem "empire" disappeared almost as quickly as it emerged. Its achievement of social and intellectual splendor appears not to be a periodic phenomenon, though the cycling of history is a most unpredictable business. By contrast, the civilizations of farther east flowed and ebbed, and their mathematical developments followed this flow and ebb.

4

FARTHER EAST—
INDIA AND CHINA

Al-Biruni roamed through much of India in the early eleventh century, in the wake of one of the periodic surges of the Moslem armies into India. Though Al-Biruni does note that he considers himself superior to the Indian in matters of mathematics and science, his description of India is looked upon as one of the better sources of information about Hindu accomplishments. For the Moslem observer was a careful, honest notetaker, acknowledging from time to time his own limitations. He was able to disassociate himself from the military conquests and was well received by the Indians.

Al-Biruni devotes much attention to the accomplishments and theories of Aryabhata and Brahmagupta, who lived in the fifth and seventh centuries respectively, but hardly mentions Mahaviracarya (Mahavira) of the ninth century. The first two mentioned were primarily astronomers, and their mathematics is directed toward the practical problems of astronomy. Pervading their deliberations are religious considerations, for religious mysticism was a most important facet of the daily life of the Hindus. Al-Biruni notes that "The science of astronomy is the most famous among them, since the affairs of their religion are in various ways connected with it. If a man wants to gain the title of an astronomer, he must not only know scientific or mathematical astronomy, but also astrology."

The matter of satisfying both science and religion has troubled many men through the ages and in various parts of the world. Evidently Brahmagupta was among those so vexed. Al-Biruni, on one occasion, notes that Brahmagupta taught two theories of the eclipses, a scientific one involving the shadows of moon and earth, and a "popular" one in which a dragon devours the luminous body.

Brahmagupta was apparently not free of professional jealousy, which also shows up from time to time even among the very objective and rational scientists of our own day. At one point Al-Biruni questions the validity of Brahmagupta's result, noting that "he is blind to this from sheer hatred of Aryabhata, whom he abuses excessively." And, again, the Moslem observer says of Brahmagupta, "He is rude enough to compare Aryabhata to a worm which, eating the wood, by chance describes certain characters in it, without understanding them and without intending to draw them. Nevertheless Al-Biruni calls Brahmagupta "certainly the most distinguished of their astronomers."

The Indians of the time of Al-Biruni believed that the earth had a globular shape with the northern half dry land, and the southern half covered with water. Al-Biruni reports that Brahmagupta believed that all heavy things were attracted to the center of the earth, thus anticipating formal statement of the theory of universal gravitation by some thousand years. Aryabhata is credited with maintaining that the earth moved, with the heaven at rest. Apparently Brahmagupta agreed, but Al-Biruni objected and in another book, *Key of Astronomy*, went to considerable lengths to refute the idea.

The interest in astronomy led the Indian mathematicians to develop trigonometry and especially spherical trigonometry to a high degree. Consideration of problems involving planetary orbits and questions related to the calendar (which Al-Biruni

Farther East—India and China

discusses at some length) prompted investigation of linear indeterminate equations. These are equations such as

$$23x - 41y = 3,$$

in which you need solutions having both x and y whole numbers.

But, in general, mathematics was subservient to religious considerations. Al-Biruni notes that the works of another astronomer, Paulisa, had not yet been translated into Arabic "because in his mathematical problems there is an evident religious and theological tendency."

Mahavira, the ninth-century mathematician, says that mathematics was given the status of one of the four *anuyogas*, "which were the auxiliary sciences, the study of which helped the aspirant to the attainment of soul-liberation."

Mahavira also listed the characteristics expected of those who would study mathematics—there were eight qualities in all: "a quick method of working; forethought as to whether a desirable result will be produced or an undesirable result arrived at; freedom from dullness; correct comprehension; power of retention; capability of devising new means of working; an ability for getting at those numbers which made [unknown] quantities known."

Early Indian writing was characterized by an almost poetical style, and this was carried over into the statement of mathematical results and problems. In fact, the mathematical writers were verbose. Al-Biruni dismisses the flowery presentations as "mere nonsense . . . a means of keeping people in the dark and throwing an air of mystery about the subject." Professor Florian Cajori, a twentieth-century historian of mathematics, remarks that "The Indians were in the habit of putting into verse all mathematical results they obtained, and of clothing them in obscure and mystic language which, though well adapted to aid the memory of him who has already understood the subject, was often unintelligible to the uninitiated."

Here is an example:

"A powerful, unvanquished excellent black snake, which is 32 hastas in length, enters into a hole [at the rate of] 7½ angulas in 5⁄14 of a day; and in the course of ¼ of a day its tail grows by 2¾ of an angula. Oh ornament of arithmeticians, tell me by what time this same [serpent] enters fully into the hole."

To facilitate the framing of verses containing mathematical problems and arithmetic rules, Indian writers used a variety of descriptive words to refer to numbers. For example, zero was called "point" or "heaven" or "sky." "The beginning," "moon," and other words meant "one." "Four" was synonymous with "veda," the sacred code which had four parts, or with the word which meant the four cardinal points, or with "oceans." This made it easier for the aspiring arithmeticians of the time, but certainly caused confusion among latter-day observers and translators.

Another characteristic of Indian mathematical works of this era was the absence of proofs or explanations. They stated the result, the naked theorem, and no more. Al-Biruni complained of this: "We give the just-mentioned calculation of Brahmagupta, simply reproducing his words without any responsibility of our own, for he has not explained on what reason it rests."

The classic, and extreme, example of this neglect of explanation is attributed to Bhaskara, a notable twelfth-century Indian mathematician. He presented a diagram that demonstrated the right triangle relationship and explained, "Behold!"

Bhaskara, by the way, being addicted, as indeed were most others of his time, to astrology, consulted an astrologer about the coming marriage of his daughter, Lilavati. He was told that the stars indicated great misfortune for the girl unless she married at a certain hour on a certain day. On the specified day the bride-to-be watched as the level in her water clock told of the approach of the propitious hour. But a pearl fell from her necklace and blocked the hole in the clock, stopping the flow of water. Thus the right moment for the marriage passed unnoticed, and of course, she could not defy the stars by marrying at some other time. As a consolation, Bhaskara decided to name his great mathematical work for his daughter, but many people have regarded this as a poor substitute for a happy marriage.

There is a great danger in this practice of ignoring the need for explanations—the same danger that attends unembellished answers on homework. The work is suspected of not being original. In particular, a strong argument can be made for the theory that many Indian mathematical results are merely reproductions of earlier Chinese efforts. For example, consider the diagram showing the right triangle relationship and dating possibly to the time of Pythagoras, certainly not later than about A.D. 180. Compare it to the Bhaskara diagram. This appeared in an ancient Chinese mathematical work called *The Arithmetical Classic of the Gnomon and the Circular Paths of Heaven*, so you can see that the Indians did not have a monopoly on picturesque mathematical speech. It is but one of many such exam-

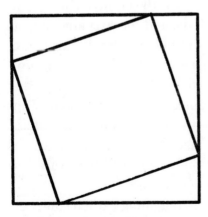

Early Chinese Diagram

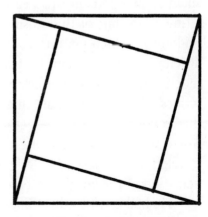

Bhaskara's "Proof"

ples of Chinese mathematical results which antedate very similar Indian results.

You can argue for independent development of ideas. Indeed this happened with developments in widely separated cultures, and even in locales not so far apart—the Leibniz and Newton independent development of the calculus, for example. But in the case of the Chinese and the Indians there is a plausible explanation of transmission of results.

Buddhism, which had originated in India, was introduced into China about A.D. 75, probably by wandering monks. The philosophy of Buddha thrived in China, and through visiting Buddhist monks, ideas were exchanged between the two cultures. Chinese monks, in particular, made the long trek to India to study at the monasteries and university centers there. Reports written by some of these travelers have been preserved and provide quite reliable accounts of what went on in India at the time. Their principal interests were in the social and cultural climate, philosophic and religious ideas, and there is little in their writing about Indian mathematics. In fact, it has been suggested that these monks exchanged Chinese mathematics for Indian metaphysics.

Fa Hsien, one of the most renowned of these travelers, spent several years in India early in the fifth century. This was at a time of decline of Buddhism in India and the strengthening of Hinduism, which permitted, if it did not specifically favor, the developing caste system. Fa Hsien's account indicates that "untouchability" was firmly established by that time. He also described the free hospitals, maintained by well-to-do gentry in the important towns and cities, as well as animal hospitals. There were resthouses along the main thoroughfares at which travelers could have free lodging and food—a fifth-century anticipation of the youth hostel idea of our time.

He had considerable difficulty in obtaining written copies of the Buddhist texts, for paper, though known in China from

Farther East—India and China

early in the second century, was not yet much used in India. At that he nearly lost his entire collection during a typhoon on the return trip to China.

Probably the best known of the Chinese monks was Hsuan Chuang, who spent some fourteen years in India in the seventh century. This was the time of Harsha, a king who had united and brought prosperity to much of Northern India, following a period of chaos caused by invasions from the steppes of Asia by a people loosely characterized as "Huns." Hsuan Chuang notes the continued decline in the popularity of Buddhism, but it was an era of tolerance, and he was much in demand. While studying at the Buddhist university-monastery in Nalanda, Hsuan Chuang received a summons to the court of a minor king, Kumara, who was a Hindu.

He indicated a preference for remaining in Nalanda and studying, but after the third invitation was refused, Kumara promised that he would send an army, with elephants, to trample Nalanda into dust. Hsuan Chuang hastened to accept the invitation. Kumara was then asked by Harsha to send Hsuan Chuang to his court. To this request Kumara replied that Harsha might take his head, but certainly not his guest. When the royal messenger returned it was with the brief order, "Send the head at once!" Shortly thereafter, Hsuan Chuang was on his way to the court of Harsha.

His account is one of the best descriptions of Harsha, who is known as an able and benevolent ruler. Taxes of the time were not excessive, reported Hsuan Chuang, being one-sixth of the gross product of the crown lands. This money was spent as follows: one-fourth to expenses of government, one-fourth to the endowment of public servants, one-fourth to charity distributed among various religious sects, and the remaining one-fourth to the subsidy of intellectual efforts.

There is no direct evidence that these monks served as a link between Chinese and Indian mathematics. But in light of well-authenticated accounts of Chinese mathematical progress, and even accepting the latest possible dates for certain developments, the Chinese were ahead of the Indians. There are a number of social and cultural considerations involved, and perhaps it is reasonable to view the situation in China, through the perspective of a man who lived at the same time as Al-Biruni.

Dream Pool Essays, to the eleventh-century Chinese, was an appropriate description for a book summarizing academic, social, and philosophic vogues of the day. The author, Shen Kua, was truly a "Renaissance man," though he was half a millennium and half a world removed from that phenomenon which Westerners are inclined to call *the* Renaissance. He was first and foremost a scholar, but he served the government of the Sung dynasty both as an ambassador and as a military commander. For a time he was director of a hydraulic works, and later was chancellor of the Han-Lin Academy. His activities reflected the important facets of the Chinese society in the eleventh century. The book contained sections on mathematics, astronomy, the calendar and cartography, along with the more "dream poolish" divination, magic and folklore.

Elsewhere, Shen Kua described his travel methods. He gives a list of travel essentials which is revealing of the sophistication of the day—a raincoat, a chest of medicines, a box of preserved food and tea, a rhyming dictionary, and a lute. He suggests also paper, ink, scissors, candles, chessmen, and a folding chessboard. The traveler should also bring along a box for the books he will buy on the trip, and some insect powder to protect the books from worms.

China, at the time of Shen Kua, was fortunate in having had a line of emperors, the Sung dynasty, who were inclined toward matters of the intellect and the arts, rather than toward the waging of wars and dissipation, which frequently characterized the Oriental monarchs. In fact, it has been said of them that they were too civilized for the world of the eleventh century,

Farther East—India and China

and ultimately the dynasty fell before the "rude nomads of the Mongolian steppe." But this decline was well in the future for Shen Kua.

He could look back on a long development of Chinese mathematics, described in terms of a continuous tradition of written language. Probably because the written language is ideographic—words and phrases are represented by what amount to small pictures—it changed practically not at all from the "oracle bones" of the fourteenth century. This made easier the charting of the course of the development of mathematical techniques and ideas, among other matters. The language is related to another important facet of Chinese mathematical tradition, the co-ordinate system. The Chinese had not developed the geometric notion of curves to be related to such co-ordinate systems, as had Apollonius and his compatriots. But they certainly recognized the usefulness of such a scheme in showing the relationship between two variables.

The contribution of the Chinese to the development of our modern numerical notation is probably slight, though their ideographic written language must have led them, and may have influenced the Indians, to use the same single symbols over and over in a positional system. They did use a decimal scale of notation, possibly as early as the fourteenth century B.C., and the zero was in common use as a numeral certainly by the time of Shen Kua, and probably much before. Calculating was done, mainly, with counting rods on boards. This device is an early "digital computer," antedating another, and better known, digital computer, the abacus.

In light of this, perhaps, it is not unexpected that for many developments in Chinese mathematics there are present-day analogues that have computer applications and implications.

Our "electronic era" brings a remarkable combination of sophistication and simplicity. The computer itself is an extremely complicated piece of equipment, and much of the notation and format for getting the problems into the computer is new and relatively sophisticated. On the other hand, the basic operation of a scientific-type computer, that is, the way it does problems, is quite simple. The computer adds very rapidly.

The mathematics for computers reflects this ultimately rather simple numerical basis. It is, in fact, usually called "numerical analysis" or "numerical methods." Among these "numerical methods" are some that are very much like techniques developed by the Chinese before the decline of the Sung dynasty. Reasons for their development are about the same as for the recent surge of interest in "numerical analysis"—the Chinese needed mathematics suitable for the rods on the counting board.

One important computer technique involves, as input, a first approximation—a guess at the solution. The computer uses this approximation to generate a new and better approximation, which it tests against the original conditions. Finding this approximation unsatisfactory, it continues to generate numbers until the conditions are satisfied, at least within a specified margin of error. Of course, the computer can cycle through several hundred or even several thousand approximations in the time it takes to tell about one.

The Chinese historical analogue to this technique was later called in Europe "The Rule of False Position." They applied it mainly to the solution of linear equations in one variable, such as:

$$3.2x - 1.8 = 0$$

Generally the Chinese as far back as the first century began with two guesses ("double false position") and manipulated those to arrive at a satisfactory value for x.

Now this certainly seems a hard way to go at so simple a matter, but remember, there was a real problem with notation. At that, this may not be a bad way to have beginning students solve equations. The technique was transmitted to Europe, via the Moslem writers, and found favor there as late as the sixteenth century. The Chinese, incidentally, referred to the false position device by their phrases for full moon and new moon—"too much" or "too little."

Books on numerical methods these days contain sections devoted to "interpolation techniques." Such devices facilitate the writing of formulas from collected data. Chinese mathematicians during the Sung dynasty developed similar procedures.

The general approach is particularly useful in filling in values in tables for which straight generation of data is impractical, and hence, the devices are referred to as "interpolation formulas."

The Chinese, at least as early as Chu Shih-Chieh (early fourteenth century) were using what amounts to matrix notation, which mathematicians and scientists of our time find very useful in dealing with computers.

The Chinese were concerned with what we now call Diophantine problems—that is, there are more variables than equations—but they require a solution in integers.

Within the last few years, researchers in computer design have applied an idea conceived by the Chinese of about a thousand years ago. This involves a way of representing numbers, a variation on the remainder system, which is usually called "modulus representation."

In India the prevailing philosophic outlook was mystic. The deeper thinkers sought "liberation of the soul" and "oneness with the Absolute." The less intellectual were concerned about the effect of the stars and planets on their daily lives. The former related not at all to the development of mathematics, and the latter fostered developments in mathematical astronomy. But in China, there was a convenient balance between two contrasting philosophic schools—the Taoists and the

Confucianists—and both contributed to an atmosphere in which mathematics flourished.

The Taoists are generally described as mystics—they "walked outside society." They were, in fact, firmly opposed to the feudal system, which was the rule of the day, and their withdrawal from society reflected this protest. On the other hand, they were committed to the idea of careful scientific observation and keeping of records. To them "knowledge" meant "natural knowledge," and this lent itself to mathematical description.

The latter-day disciples of Confucius, by contrast, were very much concerned with "social knowledge." They advocated strong central government and a well-ordered social structure. This kind of rationalistic atmosphere was quite conducive to the development of mathematics and science as tools for what we would now call "social planners."

Since the Taoists were not participating in government, the social structure of China showed the influence of the Confucianists. In particular, there was a significant bureaucracy, to which one gained admission and promotion via examination. There was considerable government activity in many areas, particularly agriculture and the military. Both required practical mathematics.

China, at the time, was an agricultural nation, so there was much concern for irrigation projects, building granaries and the devising of an accurate calendar. Much accounting needed to be done, and there was the inevitable (it seems) business of levying and collecting taxes. Such matters of practicality—and the bureaucracy was a long established tradition by the time of Shen Kua—were certainly factors in the development of arithmetic and algebra, in contrast to the abstract geometry of the Greeks.

Apparently there was some concern for the behavior of these bureaucrats, as indeed there is in our time. Astronomer

Farther East—India and China

and writer Shu Hsi writes in the third century about the tax collector:

"The winning of his favor depends upon [gifts of] rich meat, and the securing of his support upon good wine. When the harvest work is finished and the levies are to be made, and he gathers the head men of the village and summons the chiefs of the hamlets to register holdings and names . . . then chickens and pigs fight their way to him and bottles and containers of wine arrive from all directions. Then it is that a 'one' can become a 'ten' and a 'five' a 'two.' I suppose this is because hot food is twisting his belly, and the wine-god obstructing his stomach."

Apparently a mere twist of a rod would transform a "one" into a "ten," or vice versa.

The needs of the military ran to such things as surveying and indirect measure. The *Sea Island Arithmetic*, as early as the third century, gives many applications of the right triangle to the measurement of the depth of a ravine, the height of a tower on a plain, the size of a city seen from a mountain. The military, the bureaucrats, and the many travelers of the time also needed maps, and the Chinese excelled in this business of mathematical cartography.

The tradition of Chinese map making dates to the third century. Cartography of that era could hardly be called "mathematical," but the tradition was established. One of the first map makers to be mentioned by name was a woman, Feng Lao, who apparently was well thought of in her time for a variety of talents. She also gained reputation as a historian, calligrapher, and businesswoman, and was entrusted by the emperor with important diplomatic missions.

In the third century another lady cartographer, sister of the prime minister, attracted the attention of the emperor by proposing that regions be outlined in embroidery, rather than painted over. Most of the maps were drawn on silk, which may account for the prominence of the ladies among cartographers. Silk cloth made a very convenient medium for maps, since the position of a place could be fixed by following a warp and a weft thread through the location to appropriate scale along the edges of the map. Here again you see evidence of the Chinese application of a co-ordinate system.

A male contemporary of the lady just mentioned was one Phei Hsui, who has been called "the father of scientific cartography in China." He has, in fact, been compared with Ptolemy, though none of his maps survive. His reputation rests upon a book in which he enumerated the principles of map making. These are six, according to this third-century scholar:

1. The proper selection of graduated divisions—that is, the scale to which the map is to be drawn
2. The laying out of a grid consisting of parallel lines in two dimensions, which is the correct way of depicting the correct relations between the various parts of the map
3. The pacing out of right-angled triangles such that the third side shows the distance between places—the hypotenuse, with the two "legs" along lines of the co-ordinate network
4. Proper measurement of the high and the low
5. Careful measurement of right angles and acute angles
6. Measurement of curves and straight lines

(The last three principles, notes Phei Hsui, "are the means by which one reduces what are really plains and hills to distances on a plane surface.")

This interest in applying mathematical methods to map making was continued by the Chinese through the "Middle Ages" and to modern times. One particularly remarkable effort was a "Map of both Chinese and Barbarian Peoples within the [Four] Seas," completed in 801 by Chia Tan, and a twelfth-century "Map of the Tracks of Yu the Great," carved in stone, and surviving to the present. Chia Tan also prepared maps of itineraries to Korea, Tonking, India, Bagdad.

Farther East—India and China

The Moslems did much to advance the art of cartography, both through their study of the early contributions of Ptolemy and other geographers, and through charts and records kept by their own seafarers, who probably picked up techniques through their contacts with the Chinese.

Mathematicians and nonmathematicians of every culture have been interested in finding a relationship between the diameter and circumference of a circle, that remarkable number that we call "pi." The Chinese are no exception. Serious interest is indicated by their early attempts to square the circle . . . the only one of the three "classical constructions" that the Chinese have recorded work on.

Archimedes used inscribed and circumscribed polygons of

96 sides as a basis for saying that pi was between $3^{10}/_{71}$ and $3^{10}/_{70}$. Liu Hui, about the year 255, topped this effort by using a polygon of 3,072 sides. He gave the ratio as approximately 3.14159 and noted that he could do better if necessary.

This might seem to be about as good an approximation as might be needed, even fifteen hundred years ago, but a fifth-century mathematician made a further improvement, placing pi between 3.1415926 and 3.1415927.

This same scholar, Tsu Chung-Chihl also used $355/_{113}$ as an approximation to pi. This figure has been of particular interest since it is a term in the continued fraction expansion of pi. That is, pi can be represented as:

$$3 + \cfrac{1}{7 + \cfrac{1}{15 + \cfrac{1}{1 + \cfrac{1}{292 + \cfrac{1}{1 + \dots}}}}}$$

Now, if you cut off this expansion just before the $1/_{292}$, and this is certainly a propitious place to cut, you have $355/_{113}$, or 3.1415929203. Neither of these approximations of Tsu's was matched in Europe until late in the sixteenth century.

Noted earlier was Omar Khayyam's apparent discovery of that array of numbers which we now call "Pascal's Triangle." The same array, which contains the coefficients of the binomial expansions, appeared in the *Precious Mirror of the Four Elements*, written by Chu Shih-Chieh in 1303. Chu refers to the triangle as a "Diagram of the Old Method for finding Eighth and Lower Powers," implying that the pattern was recognized well before his time. Of course, this still is some 250 years before the time of Blaise Pascal, whose name is usually attached to the array, but this oversight has one redeeming feature: it is a

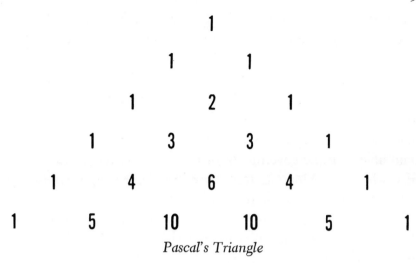

Pascal's Triangle

much more convenient description than "Chu Shih-Chieh's Triangle."

Chu, incidentally, was one of the greatest of the Chinese mathematicians. He was apparently a wandering scholar who earned his living teaching the mathematical arts. An admirer writing a preface to his *Precious Mirror of the Four Elements* notes that eager students "came like clouds from the four quarters."

The strength of the Sung dynasty waned in the thirteenth century, and this led to internal wars among the kingdoms of China, at a time when, ironically, several outstanding mathematicians flourished. Quite apart from the fact that these civil wars diverted attention and resources from scientific interests, these leading scholars were unable to get together and pool their efforts in the various academic fields. Parallel developments and duplication of efforts were natural consequences.

In the wings stood the Mongols, ready to take advantage of the dissension in China to enlarge their own empire. That they did conquer China did not mean an end to scholarly efforts. In fact, the Mongol leaders encouraged scholarly work.

Chu Shih-Chieh's great book, for example, is dated well after the last Sung claimant to the throne had thrown himself into the sea rather than be captured by the Mongols.

The Historical Records of the Yuen Dynasty (the Yuens controlled Northern China in the mid twelfth century) describes encounters between a leading mathematician, Li Yeh, and the Mongol Kublai Khan. Li Yeh, who had been an important official in the government of the province, retired to obscurity when the Mongols took over. Kublai Khan, then but a prince of the invaders, sent for him.

The Mongol expressed an interest in bringing scholars to his court, that he might benefit from their studies, and that they might enjoy his protection and support. Li Yeh observed that "The world is never destitute of wise men. If you will seek after them, they will be before you. If however, you don't seek, there comes no wise man at all. This is for only an obvious reason." The khan asked his advice on the governing of the state, and the account indicates that he listened and made a real effort to follow Li Yeh's admonitions, though the scholar refused to serve the Mongolians actively. Later, when Kublai Khan ascended the throne, he again sent for Li Yeh, who agreed to teach in the Han-Lin Academy.

Through this interest in academic matters these later Mongol rulers contributed to the extension of scholarly pursuits. Of more significance to the history of mathematics is the indirect contribution of the Mongols to the interchange of ideas between East and West. But first we should note the change in atmosphere in the West which makes the idea of cultural exchange meaningful.

5

THE WEST'S AWAKE

The West's awakening, its renaissance, was not a sudden happening any more than have been other great changes thus described. "Renaissance" is usually written with a capital R, and used to describe an era beginning in the fifteenth century. But the ninth century saw a "renaissance" under Charlemagne; the tenth century, the period of renewed cultural and intellectual activity under the emperor Otto. The renaissance of the twelfth century was more important in the development of mathematics and science than was the "great renaissance."

The monasteries, which kept alive the lights of scholarship through those periods frequently called the "Dark Ages," continued to flourish through the eleventh and twelfth centuries—a period of continued domination of the intellectual and cultural activities by the clergy. But they contributed also to an increased social mobility, which is an important characteristic of the awakening.

The monasteries were stopping-off places for pilgrims, both monks and pious lay people, en route to Rome. The monasteries ensured certain regular communications between widely separated establishments. With this mobility and communication came increased interchange of ideas, leading eventually to contacts with the East, particularly through the Moslems.

Then the importance of the monasteries gradually declined, as the pilgrims were joined on the road by traders, students, teachers, and jongleurs. These groups of travelers reflect the changes that attended the "awakening"—the growth of towns, the development of crafts and trades, and the decline of the feudal system.

All this was a gradual process, but let us arbitrarily assign the year 1000 as the beginning of the awakening, which became quite important in the development of mathematics, and which was, in turn, substantially affected by developments in mathematics. At this approximate date there flourished an individual who epitomized the characteristics of the next two centuries. Incidentally, he established a reputation among his contemporaries as quite a mathematician.

Gerbert of Aurillac has been described as "the redeeming intellect of the tenth century." That he became involved in the politics of his time, with all the "in-fighting" that went with it, is rather ironical, but not unexpected when you consider the kind of person he was.

Gerbert was a monk, but no stay-at-home like the Venerable Bede. After receiving such education as was afforded in the monastery at Aurillac, he was urged by the abbot of the convent to take advantage of a chance to travel to Spain. There he must have had some exposure to Moslem learning. His correspondence indicates that he was particularly interested in Arabic astronomy, and, further, that there was at the time rather widespread interest in and knowledge of the Arabic language and the science of the Moslems.

Gerbert attracted the attention of the pope, and through him was recommended to the emperor, Otto I, who engaged the young monk as a teacher of mathematics, logic, and letters. It was as a teacher at the court, and particularly at the cathedral school in Rheims, that Gerbert is best remembered. His pupils included a future emperor and young clerks who themselves became outstanding teachers.

Much of the learning from the Greek and Roman eras was preserved in monasteries. Manuscripts were copied by monks and eventually transmitted to the emerging countries of Western Europe.

He did not adopt the methods of calculation which the Moslems probably imported from India, although his long division algorithm was no more complicated than many that were used in Europe long after Moslem methods had been introduced. Gerbert worked on a board-type abacus, which forced recognition of place value. Even in the year 1000, people who had to do long division were not calculating in what we now call "Roman numerals."

Gerbert's interest in Moslem science and astronomy and Latin literature reflected the beginnings of the revival that gained momentum in the twelfth century. He was not original, particularly in his mathematical work. As his detractors point out, he did no more than master the knowledge that was available to any literate and reasonably well-situated man of his time. The facts remain that he did study a wise variety of subjects thoroughly, and that he was an inspirational teacher, thus directly affecting men who were to become important in the revival of learning.

Most of the learned men and women of this era were prolific letter writers. Gerbert was no exception, and many of his letters still exist. Among these is one, to Adelbold, a former student, in which he explains why the formula

$$\frac{s(s+1)}{2}$$

does not give the same value for the area of an equilateral triangle as does

$$\frac{s^2 \sqrt{3}}{4}$$

The argument was a bit vague; still, this kind of inquiry was a beginning.

Most influential and versatile of Gerbert's students was Fulbert, who became head of the cathedral school at Chartres.

Fulbert was "well informed" in mathematics, though disinclined toward research. He wrote church music, some of which is still used. He also tried his hand at poetry, and even a little of what might be classed as "secular music" of the day, which has prompted one latter-day commentator to describe him as, "This fundamentalist seminary pedagogue; this eleventh-century ecclesiastic man-of-affairs, Saul-like, is found among the troubadours and humanists, a not unworthy consort for Petrarch."

Fulbert was friend and adviser of King Robert the Pious, who ruled much of what is now France. This friendship was strained at times because of the behavior of Robert's queen, who "plundered church property," and must have been quite a formidable lady. Fulbert refused to join the bishop in decreeing her excommunication—not because of his friendship with the king, but, rather, on the very practical grounds that "there was no one who dared carry the excommunication to her."

As Gerbert rose in the hierarchy of the church, becoming first the Archbishop of Ravenna and finally pope (as Sylvester II), his attempts to reform and enlighten earned him the enmity of those who preferred to see things go as they had been going. He lamented in one letter that Fortune "has loaded me with most ample store of enemies. For what part of Italy has not my enemies? My strength is unequal to the strength of Italy! There is peace on this condition: if I, despoiled, submit, they cease to strike."

As seems inevitable for those who become involved in politics in any era, his own actions began to reflect intrigue and a certain amount of unscrupulousness. His successes, coupled with his continued interest in Moslem science, when churchmen later began to think about it, earned him a reputation as a magician and one who conspired with the powers of darkness. A thirteenth-century manuscript, for example, includes a statement that Gerbert became archbishop and pope with demon aid, and had a spirit enclosed in a golden head whom he consulted as to knotty problems in composing his commentaries on arithmetic.

Gerbert, because of his travels in Spain and Italy, has been called the first of the "wandering scholars" who roamed through Western Europe, teaching and seeking learning. The tradition long antedates him, however, and he did travel in a style well above that of the "Goliards" who made their contribution to the revival of the twelfth century.

Gerbert's predecessors in this role were those who traveled as early as the sixth century to the schools in Ireland, where they received food, and books for their studies and teaching, free of charge. There was a reciprocal arrangement for the Irish monks who traveled about Europe. In the eleventh and twelfth centuries the wandering scholars were principally clerks of the Church, and then, in increasing numbers, young men who might well be described as "anti-cleric."

These students were given the privileges of the clergy, including immunity of arrest. An instance is recorded of one provost who arrested and fined a traveling scholar and then was himself deprived of office and fined. They thus enjoyed the remarkable combination of practically unlimited freedom and no responsibility. Their excesses were deplorable, but they served the very necessary function of providing criticism of the Church of medieval times.

The situation of the wandering scholars held considerable attraction. One schoolmaster at St. Denis complained of the unofficial rivalry between himself and a traveling scholar who played the violin. The fiddler had apparently got around the parents of the students with his flatteries and had emptied the regular school of its students.

As the "Order of Wandering Scholars" became more secular, their criticisms and satires of the Church became more biting, and their poems more iconoclastic. Eventually, the Church cracked down. The members of the order had made their contribution to the reawakening of Europe, however. Their poetry and songs were of the spirit of the Great Renaissance, and their "guild" was like those of the guilds of students who formed the earliest universities.

The cathedral schools developed around the principal churches of the time—at Orléans, Rheims, Chartres, Lorraine. The school at Chartres was particularly well known, through the reputation of Fulbert, who stressed the liberal studies, as contrasted to concentration almost strictly on the Bible and commentaries on the Bible by the Church fathers. But the curriculum was fairly typical, consisting of the seven arts, divided into the Trivium and the Quadrivium.

Grammar, the first of the Trivium, included a study of literature. Works in the library included those of Livy, Virgil, Ovid, and Horace and also the Christian writers. The Trivium also included "rhetoric," which meant the study and imitation of the Roman and Christian orators, and "dialectic" or logic. Particular stress was placed on the works of Aristotle, with commentaries by such scholars as Gerbert and Boethius.

Boethius was the most important single source, particularly for the four mathematical studies that composed the Quadrivium. For arithmetic the students at Chartres used an abacus that had Roman numerals at the top to indicate place value—an intermediate step between the nonplace value system of the Romans and the positional notation of the Eastern people. In geometry and music Boethius was also the principal reference. Astronomy was taught as a basis for computation of the Church calendar, just as it was in the time of Bede. But some of the Moslem ideas, now third or fourth hand, must have filtered down to the students.

The curriculum at Chartres included what passed for a study of medicine, being chiefly recipes and processes commonly attributed to Hippocrates. These were arranged in verse form by Fulbert "for more convenient memorizing."

The general inclination in the cathedral schools was to regard the "liberal" arts as mere handmaidens to the study of theology. After all, these were cathedral schools preparing young men for work in the Church. Fulbert's emphasis on nontheological studies was one indication of a change. Another was the growing reputation of the school at Lorraine as a "mathematical center."

Now eleventh-century Lorraine was hardly a Princeton Institute for Advanced Studies, nor was it comparable to Uleigh Beg's Samarkand. The principal object of study, and medium for calculations, was the abacus. Students carried the knowledge of the operation of this earliest of digital computers (aside from the fingers, of course) to all parts of Europe. In England, the Lorraine techniques were quickly applied to a function for which descendants of the abacus, the digital computers, are now used—calculation of tax returns. Clerks of the Exchequer, abaci at hand, became adept at this, and records survive to this day of levies upon the wealthy for the support of the government.

And now we approach an era of intense trading. But before there can be the commercial trading that leads to this intellectual trading, there must be a basis upon which to develop commerce. In the twelfth century the guilds, first the Guild Merchant and then the craft guilds, formed this basis. With their ascendance came the evolution of towns.

Italy had a continuing tradition of urban living, dating to the days of the Roman Empire. But in other parts of Europe the Roman towns had been far apart, and not very sturdy at best, and had dwindled away as the empire crumbled. Feudalism was the order of the day, and life centered around the manor and castle in the period known as the early Middle Ages.

In the twelfth century, craftsmen, who had been in the

Pythagoras, carved on the cathedral at Chartres, which was the sight of a very early school that developed into a university. Courtesy of the New York Public Library picture collection

"employ" of the feudal lords, began to venture from the reservation and assume a degree of independence. This was an extraordinary step for the individual, with the lords of the manor controlling every facet of the social system. The rebels found it convenient to combine their efforts to secure privileges from the exactions of the feudal lords. These privilege exemptions were written into the charters of the towns.

One of the important privileges of the town was that of conducting trade, and from this privilege came the Guild Merchant, who was, in fact, part of the government of the town. The craft guilds developed on this pattern and from the apparently natural tendency of individuals with similar interests to associate and organize. These associations were forerunners of the trade unions of our time. The revival of trade, not so incidentally, produced a need for more efficient bookkeeping—hence arithmetic—and eventually brought contact with the Moslems and the East. This contact, in turn, produced a solution to the more efficient bookkeeping problem.

The towns also began attracting students in some numbers, and these youthful scholars organized themselves into guilds. From the student guilds, and the guilds of masters or professors, grew the universities. Paris, for example, was a particularly important center of trade, of guild organization, and, later in the twelfth century, a university center. Its strong field was theology and, to an extent, the "liberal arts." When the infant university featured outstanding teachers, such as Peter Abelard, it began to draw students away from the cathedral schools.

The university at Paris capitalized on the reputations in theology of several local cathedral schools. Bologna, in Italy, had a well-established reputation in law. There seems to be no clear-cut hypothesis for the rise of Oxford. It has been suggested that it was founded by a group of disgruntled students who had left the university at Paris, thus carrying on the tradition of the wandering scholar. (One rather disenchanted observer of the time noted that the students "learn theology in Paris, authors in Orleans, math in Toledo . . . and in no place decent manners.") Oxford was soon to become a center for the study of natural sciences . . . and the "new mathematics."

The students of these times, through their guilds, were a force to be reckoned with in the universities—particularly in Italy. They determined the rules of the universities, including when and where the professors would lecture, and, on occasion, threatened to migrate to another city and thus cut off the income of the learned masters. In fact, the Bologna law university did migrate to Arezzo in 1213, and to Padua in 1222.

It is hard to imagine students of our time going to such extremes. But the university of the Middle Ages was quite a different institution from that of the twentieth century. It was no more than a gathering of students and professors. Students found their own housing, meals, books, supplies, and, in general, fended for themselves, once the money from home had arrived.

Not all the activities were academic, though. A jaundiced observer spoke of the students "frequenting taverns at least as much as the lecture rooms, more capable of pronouncing judgement upon wine and women than upon a problem of divinity or logic."

Among the lessons in grammar were examples of special purpose letters. Those devoted to pleas for money were particular favorites. Here is an especially poignant example:

"To his father H., C. sends due affection. I am much obliged to you for the money you sent me. But I would have you know that I am still poor, having spent in the schools what I had, and that which recently arrived is of little help since I used it to pay some of my debts and my greater obligations still remain. Whence I beg you to send me something more. If you do not, I shall lose the books which I have pledged to the Jews and shall be compelled to return home with my work incomplete."

And, for balance, here is a response, written by one father: "I have recently discovered that you live dissolutely and slothfully, preferring license to restraint and play to work and strumming a guitar while others are at their studies, whence it happens that you have read but one volume of law while your more industrious companions have read several."

But for all the dissipation, the migrations and threatened migrations, the town and gown riots, the students were absorbing some of the "explosion of knowledge," and the tradition of university education was firmly established. There continued a strong influence of the clergy, however, and, with the increased emphasis on the Aristotelian logic adopted by the Church, the trend of the academic took a turn to the closed system called scholasticism.

The history of ideas is haunted by the ghosts of intellectual white elephants, which were, in life, nurtured on authoritative misinterpretations. Scholasticism was one of these white elephants. The Church scholars of the twelfth and thirteenth centuries devoted much time and effort to the use of Aristotle's system of logic in deriving consequences and conclusions from the assumed truths of Christianity. It seems safe to suggest that both Aristotle and Jesus would have been greatly surprised—and, hopefully, amused—at the developments.

In this system of reasoning logically from doctrines that were assumed to be true, there was no place for the observation, intuition, and experimentation that characterize modern science. If there was debate on a point, the question was resolved through reference to the writings of the Church fathers. Experimental results and theories that seemed to contradict the carefully reasoned conclusions were disregarded. One such dangerous doctrine was that of the eternity of matter. Another, better known, was the notion that the earth might not be the center of the universe.

Now, if you develop any kind of closed system—intellectual, social, economic, *et al*—admitting no new ideas, blood, or money, the system is likely to produce all kinds of strange results. In an intellectual system such as that of the scholastics,

The West's Awake

the arguments began to go in tighter and tighter circles. The history of scholasticism shows greater and greater "refinement" of arguments until you reach that penultimate question, "How many angels can stand on the head of a pin?"

That this question was seriously examined may be only very popular folklore, but consider this observation by Thomas Aquinas, chief among the scholastics: "Yet the place where the angel is need not be an indivisible point, but may be larger or smaller, as the angel wills to apply his virtues to a larger or smaller body."

Scholasticism was not a problem peculiar to Western Europe of this time. The increasing Moslem orthodoxy expounded by such men as Al-Ghazzali helped bring the decline of Moslem culture. Jewish scholasticism followed the Moslem trend and, through their increased reliance on Aristotle, the Jewish scholastics influenced Christians, who were inclined in that direction anyway. Possibly the earliest scholasticism was that of the Indian Buddhists, in the fifth century. The neo-Confucian philosophy, fashionable in eleventh-century China, also stressed close, logical reasoning from premises assumed to be true.

In the West there developed a strong countermovement—that of the experimental method. Experimental science eventually won out, and that changed the course of social and cultural development, mathematical development in particular, and shifted the scene of such developments to the West. It was certainly not a quick victory. Successors of the early scholastics forced Galileo to recant his experimentally based ideas. It was scholastic thinking that suppressed and delayed the Copernican "revolution." But these were definitely rear-guard actions, though they did not seem so at the time.

Not that scholasticism is dead, even today. One of the real ironies is that certain very popular interpretations of the "new mathematics" of our time reflect this kind of thinking. When

Pageantry and parades were important considerations in the Middle Ages, and seem to have persisted to the present. This drawing is by Johannes Scherr.

one hears two teachers arguing about whether or not a student should be given credit for "A plane is the *border* between two half-spaces" rather than the prescribed "boundary between two half-spaces," one cannot help but think that they are not far removed from the angels-on-the-head-of-a-pin argument.

The countermovement to scholasticism began well before that system had reached its full development. Gerbert made a small beginning, and this helps account for his thirteenth-century reputation as a magician and consort of demons. Substantial contributions were made by some twelfth-century English scholars, of whom Adelard of Bath was representative.

Adelard traveled extensively before settling down in England, visiting Moslem lands and spending some time in Sicily, which was becoming a meeting place of the Eastern and West-

The Astronomer *by Dürer. This appeared on the title page to* Messahalah, De scienta motus orbis, *printed in Nuremberg in 1504. Note the astrological symbols and the globe. The latter suggests Dürer's assurance that the earth is spherical.* Courtesy Dover Publications

ern cultures. His description of the state of society upon his return to England indicates that there existed then some of the same conditions of which we see examples today. In the introduction to *Natural Questions* he tells how he inquired after "the morals of our nations," and was told that "princes are violent, prelates wine-bibbers, judges mercenaries, patrons inconstant, the common men flatterers, promise-makers false, friends envious, and everyone in general ambitious."

Adelard goes on to say that he intended to ignore such "moral depravity . . . for oblivion is the only remedy for insurmountable ills." He subsequently concentrated on problems in natural science, including experiments on pneumatics. His work and ideas influenced Roger Bacon, who has often been called the first experimental scientist.

In general, Adelard appears not to have been much impressed by Moslem mathematics, though he did translate Al-Khwarizmi's astronomical tables and Euclid's *Elements*. He wrote treatises on the abacus and the astrolabe, but his authorities apparently were Gerbert and Boethius. Al-Khwarizmi's *Algebra* was translated from the Arabic about this time, but the translator probably was Robert of Chester.

Adelard may very well have ascribed some of his own ideas to the Moslems, for he notes that a new idea seemed to have a better chance of being accepted if it was the work of a Greek, Roman, or an Arab. If many men of the time shared this attitude, it could account for a suspiciously large number of works of the later Middle Ages which are claimed to be translations of "lost originals" of Greek or Moslem authors. Thus you will see references to "pseudo-Boethius," "pseudo-Aristotle," and others, which add to the problems of sorting out "who did what."

For all the "pseudos," doubts, omissions of these twelfth-century scholars, they made a beginning for the development of Western science and mathematics. The stage was set for reception and appreciation of the work of Eastern scientists and mathematicians. In particular, the West needed the convenient mathematical notation that is usually described as "Hindu-Arabic numerals."

6

WEST MEETS EAST

The history of civilization is a history of the transmission of ideas. This transmission happens in many ways. There is the direct handing down, from one generation to another, as in the case of the Greeks, in the "modern" West, or within any single culture. There is the partial or complete assimilation of one culture by another, as in the case of the Moslems, who were in a position to receive Greek mathematics and science by having conquered much of the Greek "world." There are instances of small-scale interchange of ideas, such as the suggested trading of Chinese mathematics for Indian metaphysics through traveling priests, which had a widespread effect. And there have been massive confrontations of two cultures.

It is such a confrontation that reached a peak in the twelfth century, having a tremendous effect on the development of mathematics and science. The most important facet of this meeting of cultures involved the Christian West with Moslem Near East, but the Far East contributed, as did the Byzantine Christians and others. Pinpointing the transmission of particular ideas is almost impossible, but one can get an impression of the general climate in which mathematical ideas were exchanged.

A study of this East-West meeting turns up unexpected

participants, in addition to scholars like Al-Biruni. There were wandering Jews, whose contributions included the translation of Eastern folk tales; kings of Sicily, described as "baptized sultans," who sponsored important translations and even contributed personal investigations; a Mongol emperor who has been called "the cruelest man in history"; Christian crusaders who may be counted a negative effect because of some of their deeds, including the sack of Constantinople; as well as many, many nameless, historically inconspicuous merchants, teachers, translators, soldiers, and pilgrims.

Many of these travelers were Jews. One motivation for travel came from their deep concern for preserving the traditions of their religion. Remote Christian communities, particularly before the founding of monastic orders, developed in isolation, or semi-isolation, which brought many changes in doctrine and forms of worship, leading to what the Church fathers called "heresies." But the Jews maintained close contact, via letters and messengers, between remote synagogues.

The Jews flourished in Moslem countries, frequently as translators of mathematical and astronomical writings, though they contributed little in the way of creative effort. At least, so it would seem now, but we must keep in mind the distortions that centuries of prejudice may have created. Since the era of prejudice against them in Christian countries had not then begun, they must be counted a significant factor in this East-West transmission of ideas.

One of the first whose name is recorded, though only as "Isaac the Jew," traveled as emissary from Charlemagne to the Caliph Harun al-Rashid, in Bagdad. His two-year venture revived a connection between Frankish and Persian synagogues, as evidenced in subsequent frequent mention of the Franks in decisions by Eastern rabbis.

In the ninth century Jewish merchants known as Radhanites traveled regularly between China and Provence.

Their route was part by land and part by sea, with intermediate stops in Damascus, Oman, Hindustan, and the Crimea. These travelers took eastward such commodities as eunuchs, girl slaves, boys, brocades, furs and swords. On the return trips they carried musk, cinnamon, camphor, medicinal herbs, probably porcelain, as well as the idea for an efficient collar harness and possibly know-how for deep well drilling.

Some three hundred years later, the more famous Abraham ibn Ezra, a native of Spain, visited Italy, France, Syria, Mesopotamia, North Africa, and, possibly, England. No proper account of his travels has been found, which is particularly unfortunate since he was much interested in mathematics. He did translate into Hebrew Al-Biruni's book on the astronomical tables of Al-Khwarizmi.

If "Rabbi ben Ezra" did reach England, he was there at the same time as a converted Spanish Jew whose activities are better documented. This man is known by his Christian name —one or another variation on Peter Alphonse—called after the Apostle Peter and "the glorious emperor of Spain who was my spiritual father and who received me at the baptismal font." Peter Alphonse apparently served as physician to the English king, adviser to Adelard of Bath, and translator of a great collection of Oriental tales. The collection is known as *Disciplina Clericalis*. It enjoyed great popularity and wide circulation in the West in the twelfth, thirteenth, fourteenth, and fifteenth centuries, with editions in Latin, French, Icelandic, Italian, German, Spanish, and English. Peter Alphonse evidently used Roman numerals—at least, the Middle English version of his collection indicates this.

Another Jew, Benjamin of Tudela, traveled widely later in the twelfth century and wrote an extensive account of "things seen" and "things heard." At this time Jews still enjoyed positions of importance in many courts, including that of the pope, and thus Benjamin had access to important sources. But the

The northern hemisphere of the Celestial Globe, done by Dürer for Johann Stabius, who in 1497 took up residence in Vienna as professor of mathematics and court astronomer to the emperor. Note the heavy astrological overlay in the woodcut. In the corners of the woodcut are four astronomers, Aratus Cilis, Ptolomaeus Aegyptus (Claudius Ptolemy), Manilius Romanus, and Azophic Arabus. Courtesy Dover Publications

value of his account was obscured in the wave of religious prejudice which dominated minds and literature of Europe for centuries thereafter.

Benjamin's itinerary took him to many of the Greek and Moslem mathematical centers—Samos, Rhodes, Alexandria, Cairo, Bagdad—and to Sicily and Constantinople, two of the most important centers of the West-East confrontation. He attended the sports events in Constantinople commemorating the birthday of Jesus, and noted that "Representatives of every nation under Heaven might then be seen gathered together." In Cairo he was a little more specific, mentioning merchants from Spain, Abyssinia, Syria, and India.

Among the Jewish contributors to the East-West interchange were an unlikely group—the Khazars of Crimea and the lower Don and Volga valleys. A nomad people, they had turned to commerce, and in the eighth century to Judaism, as a result of an application by their Prince Bulan, of what would now be called "games theory." The prince reasoned that Judaism was regarded as the second best form of belief by both Christians and Moslems, and thus might be regarded as a way of gaining maximum favor for his people.

The Khazars maintained trade with China and with Byzantium. They were even called "the Venetians of the Euxine and the Caspian." Moslem reports of the tenth century indicate that Chinese was understood in Khazaria at that time. Perhaps not so incidentally, Al-Khwarizmi represented the caliph in Khazaria for five years in the mid-ninth century.

The legendary adventures of Sindbad the Sailor have long made popular tales in many countries. Development of the legend was influenced by Greek mythology, Indian tales of "The Seven Sages," and Persian tradition, but Sindbad's exploits particularly reflect the accounts of Moslem adventurers of the ninth and tenth centuries. Two, by name Sulaiman (sometimes Soleyman) the Merchant and Ibn Vahab, left extensive accounts of their voyages, and in the Sindbad stories you can recognize actual places and people that these men saw.

They gave careful reports of the habits, governments, religions, social customs, and natural characteristics of the Indians and Chinese. Such artifacts as bells, tea, porcelain, and paintings are mentioned specifically. While mathematics is not mentioned, it seems quite likely that such merchants might recognize the superiority (for bookkeeping) of the Indian and Chinese system of numeration over that of the Greek-Roman-Hebrew tradition. This, even though they did their calculating on the abacus or counting board.

Sulaiman and Ibn Vahab reported on the water routes to India and China. But the most important throughout the ages were the land routes, several variations on the "Old Silk Road." The silk trade, supplemented by dealings for spices, was important in Ptolemy's time (he mentioned the "Stone Tower" near Kashgar) and continued so for many centuries.

Perhaps the most important traveler, from a mathematical consideration, was Al-Biruni. His edition of Al-Khwarizmi's work was very important in the transmission of the new system of numeration to Europe.

The direct or indirect involvement in mathematics by religious establishments has been noted several times. In the Egyptian, Babylonian, and apparently Mayan civilizations, the priests were the principal users of mathematics. In Alexandria, the Christians strongly opposed the pagan Greek philosophers and mathematicians of the academy there, finally burning the libraries and killing Hypatia. In the early Middle Ages in the West the monasteries were the centers of what passed for "learning" at the time, and such monks as Bede and Alcuin at least kept mathematics alive. Gerbert, of course, was a prominent churchman, as were most other Western mathematicians of the era. The entire Moslem world was religion-oriented.

It is not surprising, then, that there should be a religious aspect, or several of them, to the East-West meeting. As in the past, not all of the effects were positive.

During the "Middle Ages" the Eastern Christians, usually of the Greek Orthodox Church, were frequently in conflict with the Western Church and the pope. The Orthodox Church, centered at Constantinople, was itself a blend of East and West. There was more concern for "mystic union with God," reflecting the ideas of the Indian mystics and certain Greek philosophers, than with the logical deduction of doctrine. In fact, an eleventh-century Orthodox bishop wrote a poem asking that Plato and Plutarch be counted as Christians, "because their doctrines were so noble."

This inclination toward mysticism would seem to make Orthodox Christianity an unlikely sponsor of the development of mathematics. On the other hand, the Eastern Church never inflicted on its members the confinement of scholasticism.

The Eastern Christians were great missionaries and made important contacts in the Near East. Generally speaking, they were on friendly terms with the Moslems. Certain caliphs and Eastern emperors sponsored friendly debates between their followers. In the thirteenth century one patriarch, the head of the Orthodox Church, even allowed a Turkish sultan to bathe in a bath belonging to the church. He ordered a monk to administer communion to the sultan's children, though they very likely had not even been properly baptized.

On the other hand, the Eastern Christians sponsored periodic persecution of the Jews. Jews were not permitted to live in Constantinople. Benjamin of Tudela reported that, except for Jewish physicians attending the emperor, Jews were not permitted to ride horses in the Byzantine Empire. Thus, while Jews served in the West as translators and transmitters of mathematics and science, they made no contribution in the Byzantine Empire.

In general, then, the Byzantine Church was itself less an

active agent of intellectual development than was the Roman Church. But, by virtue of their geographic situation and the more tolerant views of non-Christian peoples, the Eastern Christians were more important in this East-to-West transmission of culture.

Several Christian sects, banished by the Church as being heretical, moved East, becoming a factor in the West-East contact. Probably the most important were the followers of Nestorius, a fifth-century patriarch of Constantinople. The point of their "heresy" seems a bit obscure now, particularly in light of recent concessions by the Roman Church. But heretics they were, going first to Syria, Armenia, and Persia and later sending missionaries into India and China. Marco Polo reported that there were Nestorian churches all along the trade routes from Bagdad to Peking.

The missionary efforts of the Nestorian Christians are at least partly responsible for the Prester John legend, which in turn stirred official Church interest in the East. According to mid-twelfth-century rumors, a great Christian prince in Asia (official and unofficial Church geography was very vague, at best), one Prester John was soundly trouncing the Moslems. Both Roman and Byzantine Church leaders made substantial efforts to contact him, particularly after a letter from Prester John reached the Byzantine Emperor Manuel. (At one point

---

Religious considerations were a most significant part of Western culture in the Middle Ages. And "religious considerations" meant "Christianity," as defined by the Catholic theologians. Dürer used religious themes in many of his woodcuts, though, I suspect, in this one, opposite, he was particularly interested in the perspective features of the scene. Contrast the expressions on the faces of Mary and the Wise Man, as the baby dips into the treasure chest in The Adoration of the Magi. *Courtesy Dover Publications*

late in the twelfth century, Pope Alexander III is supposed to have sent a letter to Prester John, or some Eastern potentate with whom he had been confused, but by this time the great prince may have been consigned to Ethiopia.)

Origins of the Prester John legend have long been debated. His prototype may have been a Mongol leader, Gur Khan or Ung Khan, who was converted to Christianity by Nestorians. The stories did stimulate interest in the East, and revealed surprising misconceptions about the East by Westerners. Mixed in with early Prester John stories were accounts of strange beasts and serpents, potent herbs, a fountain of youth, absence of poverty, crime, and falsehood.

About this same time the Mongols were becoming a very uncomfortable fact of Western life, and another pope sent an emissary specifically to them. The Church's representative, John of Carpini, was directed to protest the Mongol invasion of Christian lands and to gain information regarding the hordes and their intentions. It seems very likely that the practical pope was also looking for a possible ally against the forces of Islam.

Anyway, Carpini traveled from Lyons to Kanev, near Kiev (conferring en route with an old friend, Wenceslaus, King of Bohemia), then to the Don-Volga region where Batu, the Western Mongol commander, was encamped. Batu ordered him on to the court of the supreme khan in Mongolia (a long jaunt for a sixty-five-year-old man), where he witnessed the election and enthronement of Kuyuk, as supreme khan.

John of Carpini's official mission came to little. Kuyuk responded arrogantly that he intended to continue the policy of conquest. But the reply, mentioned at the beginning of this section, was itself an epitome of the East-West interchange. Carpini wrote a description of the manners, religion, and history of the Mongols and also of the regions through which he had traveled. His book is regarded as the best treatment of the subject by a Christian writer of the Middle Ages.

Not long after Carpini's trip to the court of the khan, the Mongols overran parts of the Moslem empire, including Bagdad. This was hardly in keeping with the pope's designs, but it was hailed in Western Christendom. The Mongols had made an alliance with the Christian king of Armenia, and there were many Nestorian Christians in their armies.

A few years after Carpini returned, another churchman, William of Rubrouck, sought out the Mongol leaders, under orders from Louis IX (St. Louis) of France. His efforts failed, as far as policy was concerned, but he wrote engagingly about his travels—an account frequently cited by Roger Bacon.

Until very recently there has been little or no direct correlation between the activities of the military and the development of mathematics. No great mathematician has been a professional soldier (Descartes was hardly a professional), and historically the adventuring of the military has produced highly disruptive social atmospheres (frequently called wars), which are not at all conducive to quiet and orderly mathematical investigations. There are notable exceptions. The Pax Romana and the Pax Mongolica both resulted from extensive and intensive military activity. The latter is particularly important to the consideration here, since the Mongols, who instituted and maintained the "Peace," brought together Eastern Europeans, Chinese, and Indians in their empire.

Genghis Khan founded the Mongol empire. He has been described as the "cruelest man in history" on the basis of his reputation for wholesale butchery (though perhaps this estimate should be up for review now—Western Christian writers, in particular, tend to overlook some close-to-home candidates). He was a military genius and a great organizer, and he did institute law and order over a wide area, though his methods may not have been particularly acceptable. A strong civil service was formed. Roads were improved; government post stations were established. It could be said of the Mongol domains as it was said earlier of the Roman, that "a virgin could walk with her treasure unmolested from one end of the empire to the other."

The concern of his grandson Kublai Khan for cultural matters and his efforts to recruit Chinese scholars to his entourage has been noted earlier. Somewhat later, and several thousand miles removed, another Mongol khan, Hulagu, sacked Bagdad and then began a campaign to promote the advancement of astronomy in the area. As was also noted, he entrusted Moslem scholars with the construction of an observatory that was equipped with the best of instruments. An adjoining library is said to have housed 400,000 volumes. Chinese scholars participated in this venture, and an astronomical treatise of the era has both Chinese and Arabic writing on the title page. (This document has not been carefully investigated. It is presently in Paris.)

This was not the first report of Chinese scholars in Bagdad. The ninth-century Moslem physician and chemist, Al-Razi, wrote: "A Chinese scholar came to my house and remained in the town about a year. In five months he learned to speak and write Arabic, attaining indeed eloquence in speech and calligraphy in writing. When he decided to return to his country, he said to me a month or so beforehand, 'I am about to leave. I would be glad if someone would dictate to me the sixteen books of Galen before I go.' I told him that he had not sufficient time to copy more than a small part of it, but he said, 'I beg you to give me all your time until I go, and to dictate to me as rapidly as possible. You will see that I shall write faster than you can dictate.' So together with one of my students we read Galen to him as fast as we could, but he wrote still faster. We did not believe that he was getting it correctly until we made a collation and found it exact throughout. I asked him how this could be and he said, 'We have in our country a way of writing which we call shorthand, and this is what you see. When we wish to write very fast we use this style, and then afterwards transcribe it into ordinary characters at will.' But he added that an intelligent man who learns quickly cannot master this script in under twenty years."

The Crusaders are frequently credited with contributing significantly to the West-East cultural interchange. They did open a direct Syria-to-Western Europe trade route—by-passing Constantinople. But in general, they were as little concerned about cultural matters as were the first waves of Mongols, and the Crusades did not bring about the kind of stability represented in the Pax Mongolica. Armies of the Fourth Crusade dealt a serious blow to the cause of "cultural redevelopment" when they sacked Constantinople—to the delight of the Turks, who had attempted the same from time to time. Many thousands of art treasures and books were destroyed in one of the least defensible actions of military history.

Three locations on the Moslem-European frontier—Constantinople, Sicily, and Spain—were focal points for the transmission of Eastern ideas to the West. Constantinople, easternmost of the three, had always been a cultural crossroads. Its location made it a natural terminus of trade routes with China and the Far East. It was the center of the Orthodox religion. Founded in the fourth century as a Greek city, Constantinople was located on the site of the ancient Greek city originally established by Byzas, and stocked by Constantine with Greek art and Greek books. Yet it was a Roman city too, officially called "New Rome, which is Constantinople."

Throughout the period when the West was "asleep" there were periodic ebbings and revivals in Constantinople, reflecting variations in the empire's political fortunes. But there was no real collapse until the Fourth Crusade debacle. The city was able to keep the Greek heritage alive, but because of many cultural influences it became truly cosmopolitan. All life in the city centered around either the dictates of the emperor or religious observances, but Oriental influences were apparent, and the

Byzantine Church and Empire did not develop the austerity of the Western Church.

Constantinople was generally a wealthy city, but with its share of slums, squalor, and slavery, to be sure. Considerable sentiment developed against slavery, and the emperor established public orphanages, almshouses, and hospitals. Prices, profits, and hours of labor were all controlled, but, generally speaking, the controls were not harsh.

While the nobility liked to claim Roman descent, the population at large was cosmopolitan, and there was much mixing of races. At various times the emperors themselves were of Slavic, Armenian, or even Arabic descent. The court was a "eunuch's paradise," since many of the patriarchs and civil servants were eunuchs. This situation gave the emperor a ruling class he could trust and there was no problem of hereditary rights.

There was constant cultural, social, and economic interchange between Constantinople and the nearby Moslem countries. Byzantine citizens went over to Islam; Arabs became Christians—depending upon whether the emperor or the caliph offered the better opportunities.

The schools of Constantinople were highly regarded throughout the Mediterranean area at least. Venetian traders from the eighth through the eleventh centuries sent their eldest sons to complete their education there, as did the leader of the developing commercial cities of Italy. In Constantinople itself, the girls of wealthy families received about the same formal education as did the boys. Women had positions of equality with men. In particular, many empresses exerted considerable influence in the affairs of state.

Apart from the general attraction of the schools and trade involvements, there were a number of specific channels for Byzantine influence into Western Europe. Rulers of Ravenna in Italy expressly promoted the study of Byzantine art and civilization. King Ine of Wessex (late sixth, early seventh centuries) invited two Greek scholars from Athens to his court. During the renaissance sponsored by Charlemagne a eunuch went from Constantinople to Aachen to instruct members of the court in Greek. The western emperor, Otto II, married a princess from Constantinople, and the lady promptly shocked inhabitants of Germany by taking baths and wearing silks. But after her death, a nun reported a vision of her in hell, thus mollifying the indignant Germans.

Her son, Otto III, spoke Greek and imitated Byzantine customs in his court. In fact, so many Byzantine monks came to Germany under his auspices, and probably worked as craftsmen, that one bishop announced that they could stay only two nights at his hospices.

After the twelfth century, Byzantine influence in Europe declined. The renaissance opened new trading centers closer to home, of which Sicily was very prominent. Roger II of Sicily broke the silk trade monopoly that Constantinople had long held. But this was only a commercial indication of the rise of Sicily.

Sicily has an ancient heritage in mathematics. If there is such a thing as right of tradition in the cycles of history, it is only fitting and proper that, in an awakening Europe, Sicily should be a principal center of transmission of the old and acclamation of the new mathematics.

Sicily received its first Greek colonists in the eighth century B.C., and its civilization remained essentially Hellenistic. It is not that other countries did not try to control the island. In the fifth century B.C., both Phoenicians from Carthage and Persians from the East attacked and were, according to legend, defeated on the same day. Sicily was a battleground in the Carthage-Rome wars.

Mathematically, Sicily was much affected by the Pythagoreans. It is impossible to specify just where Pythagoras traveled (or, indeed, whether or not he really existed), but Plato visited the island, was imprisoned by the tyrant of Syracuse, and released through the intervention of the Pythagorean, Archytas of Tarentum. The city of Syracuse would have little claim to our notice had not Archimedes lived, worked, run down the street from his bath, and later been killed there.

Sicily retained Greek traditions through centuries of Roman rule and later conquest by the Moslems. The latter followed their usual policy of toleration of non-Moslem cultures—the three, Greek, Latin, and Moslem, coexisted in Sicily peacefully and to mutual benefit. In the eleventh century Normans conquered Sicily at about the same time that they conquered England. They continued the policy of tolerance—Greeks, Jews, and Moslems enjoyed complete freedom—to the point that the Norman kings were usually regarded with suspicion by the Roman Church.

The Normans, in fact, took very readily to the Oriental ways of Sicily. Two of the Norman kings have been described as "baptized sultans." Their court at Palermo was in the Eastern style, complete with eunuchs, harems, and luxurious inanimate trappings.

It was also a center of learning and, in particular, of translation.

Roger II, the first of these "baptized sultans," has also been called "the first modern king"—a nonfeudal, farsighted, and skilled diplomat. He supervised the writing of a book of the collected geographic knowledge of the time, known thereafter as *King Roger's Book*. He sponsored translations into Latin of the *Elements* and other works of Euclid, *Pneumatics* of Hero, and a very careful translation of the *Almagest*. These translations were all made from the Greek, but they received little notice outside the Norman kingdom, since translations from the

West Meets East

Arabic were a vogue of the time, and works of the Spanish translators were preferred.

Apparently there was serious Church opposition to the translation and publication of the pagan Greek works. In the preface to the *Almagest*, the translator (whose name has never been found out) complains that "the science of numbers and mensuration is thought entirely superfluous and useless; the whole study of astronomy [astrology?] is esteemed idolatry." The remark points up a problem that plagued the development of mathematics for centuries in Western Europe. Mathematics and astrology were considered synonymous by Church authorities and, therefore, mathematics was not respectable, to say the least. The aura of official suspicion lingered until the seventeenth century, when men such as Isaac Newton made a convincing argument for the usefulness of mathematics in describing natural phenomena other than the movements of the heavenly bodies.

After a brief decline under Roger's immediate successors, imperial authority was re-established and the intellectual center revived by his grandson, who ruled as Frederick II, the second also of the "baptized sultans," and Holy Roman emperor. Many legends have developed around the name of Frederick, encouraged probably by the Church leaders who mistrusted his admiration of Moslem culture, his unrestrained curiosity and propensity for experimentation, and his personal ambition.

At a time when churchmen were rediscovering Aristotle, and Aristotelian logic was becoming the standard for academic argumentation, Frederick was saying: "We do not follow the prince of philosophers in all respects. He rarely went out hunting with birds of prey; while we, for our part, have always loved and practiced this art . . . Aristotle repeats hearsay evidence, but certainty is never born out of gossip."

Such an observation could hardly have endeared him to the Church fathers. In fact, Frederick was excommunicated five times during his lifetime and once after his death, though several of the excommunications were officially related to his delay in leading a promised crusade to Palestine.

He finally did lead a rather uninspired crusade, but his real interests remained in the intellectual activity emanating from his court and extending his personal power. For his own military ventures in Italy he established a Moslem colony from which he drew a praetorian guard. This set a precedent in Italy for the hiring of mercenaries to do the fighting. The Moslem soldiers also contributed to the diffusion of Moslem culture throughout Italy. Frederick strengthened the bureaucracy, even bringing the famous medical school at Salerno under his close supervision.

Frederick communicated with Moslem leaders, often posing to them such questions as: Why do objects partly covered with water appear bent? Why does Canopus appear bigger when near the horizon, whereas the absence of moisture in the southern desert precludes that as an explanation? Frederick was himself a scholar. He knew six languages, mathematics, philosophy, wrote an excellent scientific analysis of falconry, and probably was the intellectual superior of most of the scholars and translators whom he attracted to his court. These included the renowned (though possibly overrated) Michael Scot, who had studied and traveled in Spain and the West, and one Master Theodore, a Greek or possibly a Jew.

Michael Scot spent considerable time in Spain before joining Frederick's court. In Spain he translated a treatise on spheres which was then used by Roger Bacon. (Bacon, however, described Scot as "ignorant of the sciences and languages" and said that most work attributed to Scot was really done by a Jew, Andrew. I suspect that Bacon's real pique may be at Scot's religious idiosyncrasies. Scot espoused the ideas of Ibn Rash (or Averrhoes), a Moslem philosopher, while Bacon was a staunch Dominican.

Anyway, Frederick thought well of Michael Scot, or so we have it on Scot's word. He reported that on one occasion the emperor asked him to calculate "the height of the starry heavens," using the tower of a certain church as a reference. Though there are obvious impracticalities involved here, Scot produced the calculation. Frederick then had the tower cut off somewhat, and casually brought Scot back to the site, suggesting that he might have erred. Scot recalculated, and noted that either the heavens were more distant or the tower had sunk a palm's breadth or less into the earth.

The emperor was much impressed, or so Scot reports.

Both Scot and Theodore were interested in astrology and mathematics. They worked with Leonardo of Pisa when he was at Frederick's court at about the time he wrote the *Liber abaci*, which was so important in introducing "Arabic" mathematical notation into Europe.

Spain was a cultural crossroads long before the formation of the great cultural network of the twelfth and thirteenth centuries. There were Spanish scholars at the court of Charlemagne. Alcuin was familiar with the seventh-century writings of Julian of Toledo. Charles the Bald sent an ambassador to the Caliph of Cordova in 864, and this practice was revived by Otto I in the late tenth century.

Otto's choice of ambassador was John of Gorze, who headed a monastery in Lorraine—long a gathering place for scholarly monks from all parts of France and occasionally from England, Ireland, and Scotland. John was particularly interested in mathematics, astronomy, and music and had previously visited Rome, Monte Cassino, Naples, and possibly Salerno, bringing back Greek manuscripts.

He continued this practice of book-hunting in Spain, where he had access to manuscripts in Arabic. John may well have been one of Gerbert's sources for such documents.

Control of portions of Spain passed back and forth from

Moslem to Christian over a number of centuries, which led to two subcultures. The Moslems who were assimilated by Christianity were called "Mudéjars," and Mudéjar architecture is regarded as a distinct type. When the Moslems had the upper hand, many Christians worked for the caliph and adopted Moslem customs and the language. These people were called Mozarabs, and there are references to the "Mozarabic liturgy" of the Christians of Toledo.

The city of Toledo was a principal Spanish center of Moslem to Christian Europe transmission. Its history epitomizes the history of Spain—founded in dim antiquity (legend has it, by Jewish refugees from Nebuchadnezzar); site of a Carthaginian trading post; a colonial capital under Roman rule. Toledo was the scene of confrontations between Arians and Roman Catholic Christians, and was a center of Moslem and Hebrew cultural and commercial activity under the rule of the Moors. When the city was finally secured by the Christians in the late eleventh century, the policy of tolerance was continued and Toledo prospered as a center for academic matters and translations.

One of the most prolific of the translators was Gerard of Cremona, whose life is known chiefly through a memorandum attached by friends to a posthumously announced translation. They note his "passion" for the *Almagest*, which he sought in Toledo and ultimately translated from the Arabic. The eulogy goes on with "beholding the abundance of books in every subject in Arabic and lack of these among the Latins, he devoted his life to translating, scorning the desires of the flesh, although he was rich in worldly goods, and paying attention to spiritual matters alone."

Gerard's translations included works of Greeks—Ptolemy, Euclid, Archimedes, Galen, Aristotle—and of Moslems, among them Al-Kindi, Al-Hazen, Avicenna, and their works in algebra, perspective, geometry, and astronomy. His *Almagest* became the "standard" translation, even though it was secondhand, via the Arabic, and fifteen years later than the Sicilian version.

Many Jews did translations from Arabic to Hebrew and from Arabic to Latin. One of these, Abraham bar Hiyya, often called Savasorda, also contributed original works on geometry, astronomy, the calendar, and the inevitable astrology. He wrote an encyclopedia, long since lost, which included much discussion of Moslem music. And he may have been one of the important channels through which Moslem music became known to the West.

Some hundred years after the time of Gerard, Spain—or at least Castile and León—were ruled by Alphonso X, who was much interested in mathematical and scientific matters. Alphonso was nicknamed "El Sabio," which is frequently translated "the wise" or "the learned." While Alphonso was well educated and concerned with many academic areas including law, literature, and music, he was an inept ruler. He is best remembered for the astronomical tables that he supervised and that were named for him. The Alphonsine Tables became well known in Europe through a Latin translation by John of Saxony (in the early fourteenth century), but the astronomy has been described as hardly superior to Ptolemy's.

Strangely enough, though the tables were prepared in the mid-thirteenth century, they were written in Roman numerals. This was some fifty years after Leonardo's *Liber abaci*, and helps to point up the durability of the cumbersome Roman notation.

Alphonso sponsored many translations from the Arabic of mathematical and scientific works. He wrote a code of laws, parts of which were still in use in Louisiana in 1819. His prose writings, in Castilian, were good enough to earn him the description of being one of the fathers of Spanish literature. His poems, which he set to music, were, and are, highly regarded. In fact, Alphonso had the attributes of that much vaunted "Renaissance man," but, alas, he was politically ambitious.

In particular, Alphonso wanted to be the emperor of that strange political convenience called the "Holy Roman Empire." To this end he spent vast amounts of Spanish money, which led to exorbitant taxes and such measures as debasing the coinage. Finally, Alphonso X, "El Sabio," was deposed by rebellion.

He had made his contribution, however, to the transmission of Eastern and Greek scientific and mathematical ideas to the West. By the time of his death, these ideas had gained quite a foothold in Europe and were growing more significant as scholars became interested in experimental science. They contributed to Europe's assumption of world leadership in these areas.

The critical mathematical consideration was still that of notation. That is, the Roman system of numeration had to be replaced. It is time to consider the introduction of a different set of numerals, as part of the "new mathematics," to which the East made a significant contribution.

7

NUMBERS ARE FOR MULTIPLYING

Suppose, through newspaper and magazine articles, speeches, and television programs, the suggestion were made that we adopt the following numerals:

! to represent	1
? for	2
$ for	3
¢ for	4
* for	5
@ for	6
⋕ for	7

And there would be a zero, though it is interesting to speculate upon the number of people who would object to its omission.

The response would probably be something as follows:

1. Most people would not give the idea a second thought.

2. Most of those who considered the proposal at all would shrug it off with the observation that we have a perfectly good set of numerals that serve nicely for their intended purpose, and why change?

3. A few, who took the proposal seriously, would object because adoption of the new numerals would mean changing all

mathematical and scientific tables, instruments, page numbers in books, and a myriad of other things.

4. A very small minority might argue for acceptance of the proposal on the grounds that the implied "base eight" scale of notation would facilitate communication with binary computers, that the 6/9 confusion would be eliminated, and various other less significant reasons.

Reactions from the eleventh through fifteenth centuries to the idea of replacing Roman numerals with the peculiar forms of 1, 2, 3, 4, etc., must have been similar to those just mentioned. Granted that it is easier to write 28 than XXVIII, but for this should the entire system be changed? On the other hand, writing "M" is much easier than writing 1,000, so it appears that things will even out.

Actually there was much more involved than just the switch from Roman to the numerals called Hindu-Arabic. But the thirteenth-century man in the street must have seen it only as that—and have been unconvinced. There is a present-day analogue in the disinclination of people in the United States to adopt the metric system of measure. And, a few years ago, there was considerable sentiment voiced for adoption of a base twelve system. A "Duodecimal Society" was formed, logarithm and trigonometric tables were published, but we continue on a decimal scale.

Perhaps, given time, some of these proposed changes, at least the switch to the metric system, may be brought about. The "new" system, reported by Severus Sebokht in the seventh century, used by Al-Khwarizmi in the ninth, imported into the West and publicized by Leonardo of Pisa in the thirteenth century, was not generally accepted in Western Europe until the sixteenth century.

As a beginning, try multiplying say XXVII by XLIX. If you consistently replace the Roman numerals by "Hindu-Arabic" numerals, say

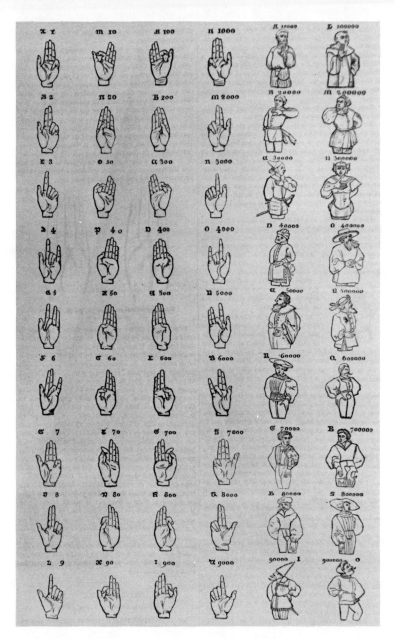

One scheme for finger reckoning was developed into a fine art during this period. In fact, a citizen who could not handle such a system competently was at a real disadvantage in the market place, in court, and probably at home when the kids hit him up for an increase in allowance.

2 for X   1 for I
5 for V   3 for L

the problem (now 22,511 times 2,312) is no easier to handle. That is, there was obviously more to the new notation than merely an introduction of different symbols. Small wonder that the Romans and most of the other people of antiquity did their multiplying on the abacus.

What, then, are the advantages of modern notation?

First, there is a convenient single symbol for each number, from one through nine. We have "7" instead of VII, "3" instead of III, and so on. These symbols are used over and over, systematically, to represent numbers, however large.

Second, in the "new" notation, the position of a digit determines its value. For example, in the numeral 24,743, the "4" between the "2" and the "7" means "four thousands," while the other "4" represents four tens. In the Roman system, say for XLIX, each "X" means "ten." It is this idea of "positional" notation that makes for easy multiplying. We happen to use a decimal positional system—positions in a numeral are related to powers of ten. But any base number will do.

My original example suggested a base-eight system, which could be positional. That is, the numeral $¢$※ would mean "three times eight to the third power, plus four times eight to the second power, plus three eights, plus seven units. The $ means one thing in the fourth place and another in the second. To do arithmetic in this system would be more amusing than instructive, but it would soon be easy, if the next condition were satisfied.

Third, we need a zero, at least as a place holder. The Romans did not need even a place holder. And they probably did not really miss having a symbol to represent the result obtained by subtracting XVIII from XVIII, or the net worth of a citizen who had been cleaned out at the races. In an efficient positional system, there must be a way to tell the difference, between, for example, thirty-two and three hundred two.

The Arabs never really claimed to have invented any of these fundamental notions of numeration. Such writers as Al-Biruni acknowledge their debt to the Hindus for the symbols. At that, the symbols that are used in the Eastern countries apparently did not originate in India.

Indian scholars make strong claims to credit for the Hindus. Considerations of national pride do affect scholarly decisions. Claims of Indian writers are particularly difficult to settle because writers in Sanskrit and other Indian languages seldom dated their manuscripts. Such claims make it difficult for the casual reader to determine just how the modern numeration system evolved. There is a similar situation in the reactions to the evidence that the Norse, among others, preceded Columbus to North America.

The Hindus may have independently rediscovered some, or all, of the basic ideas. But there are examples of their discovery and application in widely separated parts of the world. The Hindus did synthesize the ideas into an efficient system, and it is essentially that scheme which is used throughout the world today, written in some countries with marks invented by the Hindus, modified and transmitted by the Moslems.

The Egyptians discovered, at a very early date, the importance of having a single mark with which to represent a digit. Their ancient hieroglyphic system has been much publicized. It was a decimal system, even positional in a way, but highly inefficient. Some of the symbols are imaginative. Where else do you find a lotus flower playing an important role in mathematics? But what a trick it is to write five thousand seven hundred forty-six.

The Egyptians improved on this arrangement, devising a single mark for each of the first nine natural numbers and for each of the first nine integral multiples of integral powers of ten. This reform was adopted by the Greeks of perhaps about 500 B.C. They used letters of their alphabet to represent digits, and when they had exhausted the supply, began again with "alpha" and a modifying mark.

This scheme has certain advantages over the Roman numerals in the matter of multiplication, but there is the inconvenience of having so many symbols. There is the additional hazard of confusion between written numerals and written words. This early Greek system, and the improved Egyptian system before it, did permit easy representation of large numbers. Such a system requires no place holder, no lotus flowers, and no scrolls.

The Chinese, whose positional decimal system of sorts dates at least to the fourteenth century B.C., also used single symbols to represent digits. This is consistent with the nature of their written language—to represent an idea with a single, albeit often complicated, symbol. The development of a similar feature in the Indian system may reflect some Chinese influence.

Modern numerals resemble *ghubar* numbers more closely than they do the "Hindu numerals." There are two theories on the origin of ghubar numerals—that they were brought to Spain in the fifth century by neo-Pythagoreans from Greece or Italy; or that they were derived from ancient Greek-Roman symbols used on the abacus.

Much of the debate on this question hinges upon the meaning assigned to a particular word or group of words in old manuscripts. Here the philologist gets into the act. The word "ghubar" has been described as an attempt to Arabicize the Roman term "abacus," thus the second theory of origin is supported, which makes the numerals not "Arabic" at all, but rather something the Moslems picked up in Spain and that they and the Jewish mathematicians publicized. Since the Moslems usually adopted the numeration system of a country they conquered—Greek in Syria; Coptic in Egypt—they probably did the same in Spain.

The philologist, from a careful study of ancient documents, can make a case for two different words to describe Roman numeral systems. There are, they say, "digiti"—I, II, III, IV, etc.—from the idea of holding up digits. And, there are "articuli," the abacus numerals or "ghubar" numerals. On the basis of such a distinction, the argument has been raised that both Bede and Alcuin were familiar with ghubar numerals. At least, they both referred to "articuli." Even if this is a correct interpretation, neither man was aware of the advantages of the positional notation, with zero, that we associate with the use of these ghubar numerals.

Gerbert also described these numerals without the zero. But he continued to calculate on the abacus. It remained for the Hindu writers to show how their numerals, which do bear resemblances to the "ghubar," could be used in written calculations. When these calculation schemes, or algorisms, were introduced in Europe, via translations of works such as Al-Khwarizmi's, the battle between abacists and algorists began.

Elsewhere there was, if anything, greater confusion. Byzantine manuscripts from as late as the sixteenth century show four varieties of numerals: first Hebrew, positional and alphabetical, using the old Greek alphabetical symbols; second a decimal system, without a zero, a Persian modification of the Hindu system; third a positional system, but with zeros written above significant figures, as for example in 1 to mean 1,000, 21 to mean 21, 21 to mean 210, and so on; and fourth an alphabetical system with no place value feature.

There are minor complications for the translators. For example, the Moslems wrote their numerals so that place value increased from left to right. One copyist, apparently wanting to please everyone, wrote them sometimes this way and sometimes

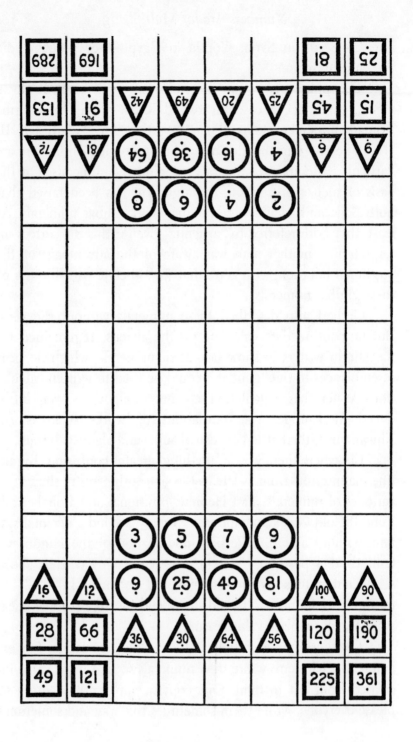

Numbers Are for Multiplying

with the increase from right to left as in modern notation. The Indians at one time used a zero to represent an unknown quantity (since zero means the missing of a number). The Hindu "five" was the same shape as our zero.

There are well-documented historical precedents, other than the example of the Chinese, for positional numeration systems. The sexigesimal (base-sixty) arrangement of the Babylonians was efficient enough, even without a zero, to permit them to do considerable algebraic and number theory work and to perform involved calculations related to their astronomy. There is an obvious disadvantage to the use of such a large base —there must be symbols for fifty-nine "digits," so the multiplication must be extended to fifty-nine times fifty-nine.

The Babylonians had no place-holder symbol, but, since numbers to three thousand six hundred could be represented with either one or two digits, they had less need than we do for such a symbol. They evidently decided whether their equivalent to 54 meant "fifty-four," "five hundred four," "five thousand four," etc., from the way it was used. They used a base-sixty analogous to our decimal fractions, and here the choice of sixty as a base paid off. Such fractional parts as $\frac{1}{30}$, $\frac{1}{20}$, $\frac{1}{15}$, $\frac{1}{12}$, $\frac{1}{10}$, $\frac{1}{6}$, $\frac{1}{5}$, $\frac{1}{4}$, $\frac{1}{3}$, $\frac{1}{2}$, and integral multiples of these, could be represented as single place decimals.

The sexagesimal notation was used extensively by the Moslems, and in some areas the system of coinage and weights

---

Opposite, rithmomachia, which was called "the medieval number game," required considerable finesse in what we would now call "the number theory." As far as I have been able to determine, the game has never enjoyed a revival in the twentieth century, and the rules are complex enough that I cannot give details here. Courtesy Scripta Mathematica

was based on "sixty"—60 falus = 1 dirham, for example. As late as the fifteenth century, Al-Kashi, a Moslem mathematician, used an arithmetic check by casting out fifty-nines.

The Mayans of Central America used two numeration systems based on twenty. One was a multiplicative system, though it was not the best for multiplying. In our own terms, this could mean having symbols for ten, hundred, thousand, and so on, and use them together with the marks for one through nine. For example, if $ meant "ten," ¢ meant "hundred," and ※ "thousand," then five thousand four hundred sixty-eight would be written 5※4¢6$8. The early Chinese decimal notation was also of this type.

The Maya also used a purely positional arrangement, though their symbols were not particularly convenient. They had a zero and probably were its original inventors.

These days, people who write and sell mathematics textbooks, and people who are inclined toward the philosophic aspects of mathematics, make much ado about the zero. It does represent a number, though a troublesome one, since it does not fit into one or another of the convenient categories. In particular, it is neither positive nor negative. It behaves badly as an exponent and in division. In general, zero is a bother.

The ancients were spared all this concern, for those who had a symbol for zero regarded it only as a means of showing an empty position in a numeral. Many mathematically sophisticated people struggled along without any such symbol at all. The Greeks sometimes used the letter omicron to show missing items, but they were inconsistent. Ptolemy, for example, used a zero only with sexagesimal fractions.

Aristotle gave some consideration to that favorite bugaboo of arithmetic students, division by zero. Aristotle evidently regarded "zero" as having the same relationship to the set of numbers as does a point to a line. In a sense, he anticipated

Numbers Are for Multiplying

some of the very rigorous work on numbers of the late nineteenth century. And, of course, this view of "zero" is in keeping with the earlier ideas of Zeno of Elea, and the general inclination of the Greeks to view mathematics in a geometric setting.

The first specific evidence of a regular and symbolic use of zero, aside from that of the Maya, came to light in Indochina at a site dating to about A.D. 900. This is not so surprising as it may sound, even if the zero was first used there. It was in Indochina that the Chinese and Indian civilizations met, since the remainder of their border was geographically forbidding. That such a mixing of ideas should produce significant innovations is not unbelievable.

The Indians may have been the first to think of zero as a number, however. Brahmagupta noted that zero times any positive or negative number equals zero. Further, that zero times zero is zero, thus emphasizing the uniqueness of the number. He expressed some doubt about the result obtained when a number is divided by zero, and argued that zero divided by zero equals zero. Both Aristotle and Brahmagupta were talking about the number itself—a philosophic problem really—and seemed little concerned about the symbol, which is so important in the everyday sense.

Bhaskara, some six centuries later, stated informally that $0 \div 0$ equals "infinity," but he did not elaborate, nor did he commit himself on the $0 \div 0$ question. These investigations by two of the leading Indian mathematicians reflect Indian concern for philosophic matters, in general, and may well represent a significant portion of original contributions from India to pure mathematics.

Bhaskara's example apparently was not followed by such prominent Western mathematicians as Leonardo, Sacrobosco, or Jordanus. The first Westerner to regard zero as a number probably was a late-fifteenth-century mathematician, Nicolas

Chuquet. But these other men may be forgiven. They were struggling just to get the "new mathematics" generally accepted without getting bogged down in details. This turned out to be quite a struggle.

The first edition of Leonardo of Pisa's book *Liber abaci* appeared in 1202. This was the first serious effort aimed at convincing the reading public, at least, that they should put aside their abaci and write out calculations, using the new mathematics. Through the centuries descriptions were published of many methods of adding, subtracting, multiplying, and dividing. Such schemes were known, and still are, as "algorisms"—a Latinization of "Al-Khwarizmi."

Some of these algorisms were given in verse form. Some contained observations on the state of mathematics in general. Little or no attention was paid to the "whys" of the business, probably because the writers themselves did not understand why the schemes produced acceptable results. But, while the explanations of the algorists may not have reflected much imagination, the methods themselves did.

Consider, for example, the multiplication method given in a twelfth-century translation of Al-Khwarizmi. The problem used for illustration was: Multiply 324 by 264:

```
  324       6       324     78324    79224
  264     2 6 4     264      264      264
```

The 264 has been multiplied by 300, from left to right. The partial product, 792, appears in the upper line, together with the digits which still must be multiplied. The 264 now is moved one place to the right, and multiplication is continued with the upper 2, of the 24.

```
  79224    83224    84424    84484
   264      264      264      264
```

Numbers Are for Multiplying

Now 264 has been multiplied by 2, actually 20, and the partial product, 528, has been added in three successive stages to the first partial product; that is, 79200 + 5280 gives the 84480, with the 8448 appearing next to the final 4 of 324. Now, 264 is moved to the right one place, and multiplication by the final 4 follows.

```
84484    85284    85524    85536    85536
  264      264      264      264      ...
```

Can you really blame a person for preferring the abacus?

The algorisms for addition and subtraction are much easier to follow, even without explanations . . . which, of course, Al-Khwarizmi omitted. Add 826 and 483:

```
826    829    909    1309
483     48      4
```

Subtract 139 from 365:

```
365    356    326    226
139     13      1
```

The algorists were not alone in committing to writing the directions for getting the right answer. Here is a translation of Gerbert's much earlier explanation of multiplication on an abacus: "When you multiply a unit with one of the tens, then each digit belongs in the column of the tens and each article in the column of the hundreds. When you multiply one of the tens with one of the tens, then the digits stand among the hundreds and the articles among the thousands. When one of the tens with one of the hundreds, then the digits stand among the thousands and the articles among the ten thousands, etc."

The "duplation" method of multiplication is one of the less obscure algorisms. It was, in fact, applied to the abacus long before Leonardo of Pisa gave the algorists their big send-off, and is a version of a method used by the ancient Egyptians. An early description of duplation was written by Rabbi Saadia Gaon, in tenth-century Spain. He was writing on inheritances, incidentally, thus putting himself at least partly in the tradition of Al-Khwarizmi.

To multiply 86 by itself, said Rabbi Gaon, by way of an example, you begin by noting that $86 \times 10 = 860$. This is very easy to show on an abacus, though a persevering algorist probably could have made it complicated. Then begin doubling results (also easily done on an abacus).

$$86 \times 10 = 860 \qquad 86 \times 40 = 3440$$
$$86 \times 20 = 1720 \qquad 86 \times 80 = 6880$$

from which he had $86 \times 86 - 6880 + 172 + 344$.

The rabbi also showed how you should multiply $61 + \frac{1}{3} + \frac{1}{9}$ by itself—$61 \times 20 + 61 \times 40 + 61$, plus "double of $\frac{1}{3}$ of 61 and double $\frac{1}{9}$ of 61," which gave him an approximation of 3775. His result, incidentally, was a better approximation than was the "exact value" given by a modern Hebrew translator.

Competition between algorists and abacists continued through the centuries. While explanations improved little, the algorisms certainly varied. Compare Al-Khwarizmi's method with the following demonstration described as "cross multiplication," by Rabbi Mordecai Comtine, writing in fifteenth-century Constantinople. The problem: 235×348

$$8 \times 5 = 40 \ldots\ldots\ldots 0$$

$$\left.\begin{array}{l} 4 \\ 8 \times 3 = 24 \\ 4 \times 5 = 20 \end{array}\right\} \ldots\ldots\ldots 8$$

$$\left.\begin{array}{l} 4 \\ 8 \times 2 = 16 \\ 3 \times 5 = 15 \\ 4 \times 3 = 12 \end{array}\right\} \ldots\ldots\ldots 7$$

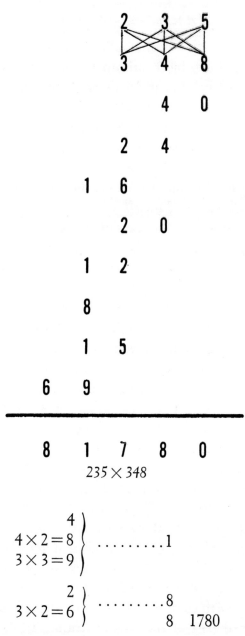

This cross-multiplication algorism is rather straightforward when you stop and think about it, but the good rabbi's "galley method" of long division is another matter. All in all, the efforts of the algorists merit the observation made by the mathematician Mizrahi, a pupil of Rabbi Comtine. "All the ancients whose work on this subject [arithmetic] reached us endeavored, it is true, to instruct the methods and processes by means of which the required results are attained; but their intention was mainly to shorten the methods leading to the results and nothing else. Their methods, therefore, looked to the students like an uninterpreted dream, because they did not understand how, nor why, the results followed from the method. They merely groped the wall like blind men."

Mysterious as were the ways of the algorists, we can hardly blame them completely for the very slow adoption of the new mathematics. An easy explanation is inertia. People tend to string along with the status quo, regardless of how attractively the alternative is presented. If you have given serious consideration to the matter of multiplying with Roman "digiti" numerals (even using duplation you may doubt the sensibilities of the good folk of the tenth through fourteenth centuries in failing to acclaim the new mathematics. But then, how many really had to write out a multiplication, or a long division, or, heaven forbid, a square root? The abacus was available for rapid calculations, and, while its operation leaves no trace of intermediate steps, most of the customers, sheriffs, tax collectors, *et al* must have been satisfied with just the answers.

At that, the Roman "digiti" system still is not dead. Elementary mathematics texts which are claimed to be "contemporary" still include sections on Roman numerals well beyond those skills needed to satisfy the clock and book chapter rationale. This is a matter of tradition now, but in the years following the twelfth-century renaissance, the ghubar numerals were considered part of the Arabic writing, while Roman numerals were thought of as part of the Latin script.

There may have been some pressure exerted by the Church. After all, there was a choice between the numerals from the center of Christendom and the symbols invented by

A page from an arithmetic book published in Venice in 1488. Some of the crisscrosses and connecting lines are reminiscent of such absurdities as are inflicted on contemporary young students of mathematics. You will note the prominence of fractions, and isn't it too bad no one thought of decimals at this stage? Think of the anguish we would have been spared. Courtesy Scripta Mathematica

the infidel Moslems, or the Jews, as many people believed. The period following publication of *Liber abaci* was one in which the Jews were persecuted in many countries of Europe, and it would not be surprising that their ideas should be dimly regarded by the persecutors.

But tradition and inertia are significant factors. Consider the attempt to have the metric system of measuring adopted in the United States. Here is a system with tremendous and obvious advantages over the assorted miscellany sometimes called the "British system." Yet now, well over 150 years after its introduction into Europe and almost 400 years after Simon Stevin proposed that coinage, weights, and measures be decimalized, some of the most "progressive" countries have failed to adopt the metric system for general use even in this scientific era of mass media.

In the centuries after Leonardo's *Liber abaci*, there was little real urgency for change to the new numeration system. Today, impetus for mathematical innovation springs from two sources—the scientists and applied mathematicians on one hand, and the "pure" mathematicians, those who are interested in mathematics for its own sake, on the other.

Experimental science was just getting a start in the thirteenth century, and the field expanded slowly, placing little demand for new mathematics. Most of the time and effort were devoted to alchemy, astrology, and astronomy. For these, there were the tables of the Moslems, and these tables were among the first manuscripts to be translated. The first Western tables, those of Alphonso the Learned, were done in Roman numerals.

By contrast today, the scientists are frequently working in areas for which there is no adequate mathematical description. The pressure is consistently on the mathematician to invent mathematics that will legitimatize the situation and provide the scientist with convenient mathematical tools by which to extend the work.

The tradition of the pure mathematician in the West was geometric. The Greeks were certainly geometrically oriented. Those who investigated what we would call problems in number theory relied even there on geometric proofs. Diophantus was the only significant exception, and he left no intellectual heirs. Much of his work has been lost.

Plato contributed to this geometric orientation with his strong words, quoted much earlier, about the respectability of compass or straightedge proofs. Even Omar Khayyam, the greatest of the Moslem mathematicians, who used the new notation, felt compelled to observe that "As the intelligent mathematicians of the past have used notations of algebraists in order to simplify the intuitive solutions, we shall also follow them. But the notations of the algebraists are not necessary. We can do just as well without them. However, with those notations, multiplication and divisions will become easier."

Thus, the notations that the Moslems passed on had few strong advocates, and it is small wonder its acceptance was so slow. The evolution through which it finally became *the* notation of the Western world, in particular, provides a background to developments in this final period of the "Age of Transmission."

8

FROM RENAISSANCE TO RENAISSANCE

The first "renaissance" mentioned in the title of this chapter is that twelfth-century affair which was really a speeding-up of the awakening of Europe. The second "renaissance" is that fifteenth-century movement, particularly in Italy, that was largely a flourishing of arts and letters. These renaissances are convenient ideas—and misleading. How can they be defined? Who is to be included and excluded? In particular, reference to renaissances implies a slack period in between, and this is far from the truth. The mathematical and scientific beginnings of the eleventh and twelfth centuries were taken up and developed remarkably by such men as Leonardo of Pisa, Richard Suiseth, Nicole Oresme, and Blasius of Parma. In the arts there were Chaucer, Dante, Petrarch, Giotto, and Uccello, whose works are highly regarded to this day.

Those who talk about the "Renaissance," and mean the Italian humanist period, have coined the phrase the "Renaissance man," by which they mean a person accomplished in a number of fields, but particularly in the arts. In this "inter-renaissance" period, you find, in addition to those just mentioned, a renaissance man such as John de Dondes (or Giovanni Dondi), who was professor of astronomy at the university in Padua, lectured on medicine in Florence, represented

Padua as Ambassador to Venice, and is best remembered in scientific circles for his remarkable astronomical clock, which took ten years to build.

This was a period of contradictions. Substantial beginnings were made in experimental science . . . that is, hypotheses were based on experimental evidence and tested against further experiments. Yet, the scholastic method was still much in favor, and questions were "settled" by reason alone. The fact was, these early experimental scientists were also Scholastics.

This was a time when the power of the Christian church was formidable. Outstanding contributors in various fields were churchmen. The Inquisition received official church sanction, and heresies were literally warred upon. But the lowly and the great continued to believe that the courses of the stars affected the fortunes of man, in general and in particular. The 1345 conjunction of Mars, Saturn, and Jupiter, for example, caused great concern and was discussed at length by these very churchmen and scientists.

This was the time of the "new mathematics," when the algorists were beginning to win converts, when Suiseth and Oresme were laying the real bases for the calculus and analytic geometry, when the painters Giotto and Uccello were developing techniques that led to the invention of projective geometry. Yet, "mathematicus" meant "astrologer."

In short, this was a lively era.

Leonardo of Pisa is more familiarly known as Fibonacci—the son of Bonaccio. His father headed a branch of a Pisa concern in Bugia, on the Barbary Coast, and did business in various Mediterranean ports, giving Leonardo opportunities to see the sights and study the cultures of Syria, Greece, and Constantinople, at least. Because of this experience the nickname *bigollo*—the traveler—was frequently attached to his name. Ironically, *bigollo* meant "blockhead" in some dialects.

Leonardo was a familiar figure at the court of Frederick II,

A TREATISE ON THE ASTROLABE · PROLOGUS

LOWIS MY SONE, I HAVE PERCEIVED wel by certeyne evidences thyn abilite to lerne sciencez touchinge noumbres and proporciouns; & as wel considere I thy bisy preyere in special to lerne the Tretis of the Astrolabie. Than, for as mechel as a philosofre seith, he wrappeth him in his frend, that condescendeth to the rightful preyers of his frend, ther-for have I geven thee a suffisaunt Astrolabie as for oure orizonte, compowned after the latitude of Oxenford; upon which, by mediacion of this litel tretis, I purpose to teche thee a certein nombre of conclusions apertening to the same instrument. I seye a certein of conclusiouns, for three causes. The furste cause is this: truste wel that alle the conclusiouns that han ben founde, or elles possibly mighten be founde in so noble an instrument as an Astrolabie, ben unknowe perfitly to any mortal man in this regioun, as I suppose. Another cause is this; that sothly, in any tretis of the Astrolabie that I have seyn, there ben some conclusions that wole nat in alle thinges performen hir bihestes; & some of hem ben to harde to thy tendre age of ten yeer to conseyve. This tretis, divided in fyve parties, wole I shewe thee under ful lighte rewles & naked wordes in English; for Latin ne canstow yit but smal, my lyte sone. But natheles, suffise to thee thise trewe conclusions in English, as wel as suffyseth to thise noble clerkes Grekes thise same conclusiouns in Greek, & to Arabiens in Arabik, and to Jewes in Ebrew, & to the Latin folk in Latin; whiche Latin folk han hem furst out of othre diverse

to whom he dedicated one of his books. Disputations between leading scholars on academic matters were encouraged by Frederick, who, as noted, liked to keep the academic pot boiling, and Leonardo engaged in at least one debate with the resident mathematician, John of Palermo. Their problems have been recorded.

> To a certain sum of money three men have proprietorship in the proportions ½, ⅓, ⅙. They divide the amount between themselves at random, not corresponding to the rights of ownership. From the amount thus received the first puts down ½, the second ⅓, and the third ⅙. The total of the amounts so put down is then divided into three equal parts and each man receives such a part. Each has then received exactly his portion of the sum of money. How large was the amount and how much did each take in the random drawing?

Leonardo had something going for him, on this problem at least, for it was contained in the algebra of Abu Kamil and other Moslem mathematicians, with whose works he was familiar. Apparently his opponent in the contest had not done his homework.

Another problem given by Leonardo in one of his books was identical, in statement and solution, to a problem written down by the Chinese mathematician Yuh Hing in 717, so you see that these little gems were well circulated.

The name Fibonacci is most often associated with a sequence of numbers which came from the following problem: How many rabbits will a pair of rabbits produce in a year if it is

---

Chaucer wrote an essay on the astrolabe (navigational and starsighting instrument) for his young son. This, opposite, is one of the first such scientific essays done in a language other than Latin. The illustration here showing the elaborate illuminated manuscript is from the Kelmscott Chaucer. Courtesy Dover Publications

supposed that each pair produces a pair each month, and the young pair produces a pair when two months old?

Leonardo argued that the numbers of pairs of rabbits would increase monthly according to the pattern:

$$1, 1, 2, 3, 5, 8, 13 \ldots$$

with each term after the second being the sum of the preceding two. This Fibonacci sequence may or may not apply to the rabbit problem, but it has apparently countless special properties, some of which are still being investigated.

Leonardo of Pisa was unique. Of the thirteenth- and fourteenth-century contributors to the development of mathematics, he was one of the very few in the West who could be called a "pure mathematician." And he was not associated with the Church.

His contemporary, Jordanus Nemorarius, a German who has been described as "second only to Fibonacci as a mathematician," was more typical of the scholar of his time. He extended some ideas on inclined planes from Aristotelian physics, though he did not apply mathematical descriptions. His mathematical works show no Moslem influence, which is not surprising, considering his remoteness from contact during the time he concentrated on mathematics and physics.

Nemorarius was a Dominican brother (he became the second general of the order, in fact) who later in life preached in Bologna and Paris, and died on the way home from the Holy Land. The Dominicans flourished under his leadership and assumed academic leadership in a number of university centers. It was to be many years, though, before they would assume the inquisitorial leadership which has tended to detract from other accomplishments of the order.

The Inquisition received its first official Church sanction during the early thirteenth century and became a factor to be reckoned with by scholars in most European countries well before the period of particular excesses in Spain. Ironically, the law code of Frederick II provided a precedent for the Church rulings on heresy and establishment of the Inquisition. At about the same time, Salimbene, a Franciscan monk who wrote a detailed description of life in the thirteenth century, was describing Frederick as "a man pestiferous and accursed, a schismatic, heretic and epicurean, who corrupted the whole earth."

Politics (and the Inquisition was hardly "nonpolitical") does indeed make strange bedfellows.

But for whatever the reason, the Inquisition was a fact of life in these centuries. The scholar did well to keep this in mind. The Jews, on the other hand, were targets not so much of the Inquisition as of the temporal rulers. Their persecution became general, culminating in periodic confiscations of properties and expulsions from various countries. Contributions of the Jews to mathematics and science declined during this period, though not necessarily because of the persecutions. Many of their leading scholars were becoming intrigued with the mystic Kabbalah, which was not immediately compatible with serious work in experimental science.

Salimbene's emphatic disapproval of Frederick reflected the general dismay over the emperor's wars against the pope and the cities that sided with the pope. This dispute was just part of the continual warfare in Italy between the various cities, of which Salimbene recounts many a gory tale. But amid this war and destruction some truly outstanding art work was done, which had important mathematical implications.

The geometry of Euclid seems satisfactory for describing the world in which we live. Parallel lines, which are everywhere the same distance apart, can be seen in ceilings and floors of rooms, tops and bottoms of windows, sides of streets. Perpendicular lines are appropriate references for corners of buildings, boxes, and other commonplace objects. A meter is a meter and a foot is a foot wherever you go—that is, a person who is six feet

From Renaissance to Renaissance

tall in London is six feet tall in New York. There are a few discrepancies in this Euclidean interpretation of the universe—triangles on a sphere do not behave quite properly according to the theorems of Euclid; the shortest distance between two points on a cylinder is not a straight line. But in general, real life manifestations of points, lines, planes, angles, etc., behave as Euclid said they should.

But if you look at a building from a distance you notice that the line of the roof and the line where two walls meet *appear* not to intersect at right angles. A six-foot person two hundred feet away from you *appears* smaller than a six-foot person one hundred feet away. If you look down a long, straight stretch of highway, the edges of the road *seem* to get closer together.

These matters of appearances began to trouble artists of the thirteenth century. Up to that time, painters had been concerned about intellectual appeal, principally to stress the themes of Christianity. Their paintings appeared two-dimensional, but appearance was secondary to the theme. If this was to be changed, and appearance to be primary, many questions needed to be answered.

At what angles should the walls and ceiling of a room be shown to meet if they were to look "right"? What is the relationship between a roof line on a drawing of a building and the lines of windows? How much smaller should a person at a hundred feet be shown, so that he seems the same height as another figure in the foreground?

One of the first to apply himself to this problem, from which evolved the science of perspective and, ultimately, projec-

---

Opposite, this study in perspective was done by Jan Vredeman de Vries, whose works summed up centuries of experimentation by artists who were also mathematicians. Courtesy Dover Publications

tive geometry, was Giotto di Bondone, a good friend of the poet Dante. Giotto's greatest works were frescoes in chapels of wealthy Tuscan families, but he was commissioned by Pope Benedict IX to do some paintings in St. Peter's. The story of the pope's initial contact with the painter has mathematical overtones.

Giotto was asked by the pope's representative to submit a sample of his work that it might be compared with samples from other painters. Giotto, "who was most courteous," according to a chronicler, took a brush dipped in red, and, with his arm against his side "in order to make a compass, with a turn of his hand he made a circle, so true in proportion and circumference that to behold it was a marvel." This, he told the courtier, was his sample.

The gentleman was taken aback, if not insulted, but he carried this artistic gem back to the pope. The latter was evidently impressed, either by the circle or by the audacity of the gesture, for he selected Giotto to do the work.

Giotto's compatriot, Dante Alighieri, is claimed by the Renaissance people as "the first humanist," but he was very much a man of his own time. In particular, he was influenced by the Greeks and the Moslems. His *Divine Comedy* has prototypes in Virgil's poems and in Moslem writings concerning the Prophet's miraculous journeys . . . and his descriptions of hell and paradise are much like those of a tenth-century Japanese bonze, Genshin.

Dante was involved in political controversies and was even exiled for an extensive period of time after one of his political treaties was burned and placed on the Index. He was interested in science, participating in a disputation at Verona in 1320 on the relative levels of land and water on the surface of the earth, and his knowledge of astronomy was rated well, for a nonastronomer. His writings contain the first reference in Western literature to the stars of the Southern Cross. He was consulted by

Giotto on selection of subjects for paintings. In short, Dante's scope of activities was far wider than that of the humanists who claim him, and his era emerges as something substantially grander than "medieval."

Interest in perspective increased in the hundred years after the time of Giotto to the point that the sixteenth-century painter and biographer Vasari could report that "Paolo Uccello [in the fourteenth century] would have been the most gracious and fanciful genius that was ever devoted to the art of painting from Giotto's day to our own, if he had labored as much at figures and animals as he labored and lost time over the details of perspective; for although these are ingenious and beautiful, yet if a man pursues them beyond measure he does nothing but waste his time, exhaust his powers, fill his mind with difficulties, and transform its fertility and readiness into sterility and constraint and render his manner, by attending more to these details than to figures, dry and angular, which all comes from a wish to examine things too minutely, not to mention that very often he becomes solitary, eccentric, melancholy, and poor, as did Paolo Uccello."

At that, Uccello's works were well thought of in his own time.

Out of the experimentation of such men as Giotto and Uccello grew a geometry in which parallel lines meet at "ideal points," circles project into eclipses, parabolas, hyperbolas; right angles become acute or obtuse, and which turns out to be quite useful to the twentieth-century mathematician, if not to the twentieth-century painter.

It is convenient to date the beginnings of Western experimental science to the court of Frederick II. Well before his time, however, a practical-minded man here and there had investigated the possibility of turning baser metals into gold. The Moslems had tested and extended the results of experiments by Hero, Archimedes, and other Greeks in pneumatics, mechanics,

and optics. In the century after Frederick's time, experimenters could be found in many places in Europe. They began to give some mathematical form to descriptions of their experiments. Yet, they were, at the same time, alchemists, astrologers, and Scholastics.

This should not be taken as a derogatory observation. Alchemy and astrology continued to be very practical considerations. Scholasticism was the way of academic life. And is the situation so different now? The historian of twentieth-century science will note that "They transmuted elements, investigated psychic phenomena . . . and examined in minute detail each facet of the real number system."

Englishman Robert Grosseteste stood out among these early experimental scientists. His protégé, Roger Bacon, reports that his mathematical background was substantial—the accumulation of "30 to 40 years study." Grosseteste was particularly interested in mirrors and magnification, and his investigation at the end of the thirteenth century led to the development of eyeglasses for nearsightedness. He argued that the earth was a sphere, noting that this was "shown both by natural reason and astronomical experience."

On the other hand, he wrote a fanciful treatise on comets, which he described as "heralds of disaster." He accepted astrology as the supreme science, directly affecting such diverse activities as the planting of crops and medical practice. He associated seven metals with the seven planets and, as a practicing alchemist, believed that metals could be transmuted.

Opposite, this Jan Vredeman de Vries study in perspective is more elaborate than those straight stairways of the earlier illustration. You will note that the date indicates Vredeman is somewhat later than the period spanned by this volume, but I'll justify including him here because his work is more closely related to that of earlier artists. Courtesy Dover Publications

A Polish physicist, Witelo, constructed a parabolic mirror, evidently recognizing that property of the parabola which we still exploit in automobile headlights, for example. Witelo also built an instrument for measuring angles of refraction of various colors in a variety of media—water, glass, etc.

Roger Bacon was and probably is the best known of the thirteenth-century scientists. Much of this reputation rests on the fact that he was vocal. He issued manifestoes denouncing blind acceptance of custom and authority. He complained bitterly, in letters, of the lack of interest in "practical mathematics." A monk, he traveled widely, including to Rome, where he attended the pope briefly, and, incidentally, met Fibonacci's pupil, Compano da Novara, who was the pope's chaplain and physician. But Bacon's scientific originality is overrated.

He did experiment with the magnet, but magnetic properties, particularly for the compass, had long been known. They had, in fact, been written about by such men as Vincent of Beauvois and Alexander Neckam. Bacon's encounters with gunpowder have been made much of, although the Chinese had long known about gunpowder, and news of such a potent substance must certainly have been reported by early travelers in the Orient. Still, Bacon, in his writings, makes a strong case for the experimental approach to science. And, more importantly for this account, called for a mathematical basis to the analysis of these experiments. "Nature," he argued, "cannot be known without mathematics."

The idea of a mathematical framework for experimental science got a real boost in the fourteenth century from the work of several Scholastics at Merton College of Oxford University. Their experiments involved uniform acceleration, with their principal result stated in modern notation:

$$S = \tfrac{1}{2} V_f t.$$

(S = distance traveled; V_f the final velocity; t the time of acceleration.) Since $V_t =$ acceleration $\times$ time, this result is equivalent to the familiar $S = \tfrac{1}{2}at^2$ ... which you will often see in the form $S = \tfrac{1}{2}gt^2$ [$g =$ acceleration due to gravity] for a free-falling object.

Of course, these fourteenth-century philosopher-scientists did not have anything like our notation. Their arguments, which they called "proofs" in true Scholastic fashion, were strictly prose affairs—no formulas, no calculations, no graphs—and they make very difficult reading. Still, these discussions hint at some insight into the question of instantaneous velocity, which is one of the basic problems to which the calculus is applied, and they certainly looked upon the relationship between two variables in a way which we today would find quite acceptable.

In fact, Wilhelm Leibniz, co-inventor with Newton of the calculus, referred to the work of one of these Merton scientists in a letter to a colleague in 1696. He suggests that his correspondent consult "the writings of Suiseth, commonly called the Calculator, who introduced Mathematics into scholastic philosophy."

Richard Suiseth, or Swineshead, is the best remembered of these scientists, although "proofs" of the Merton Theorem of Uniform Acceleration by William Heytesbury and John Dumbleton have been preserved. Suiseth was listed by Cardan, that colorful sixteenth-century mathematician, as one of the ten leading intellects of the world. (Others being Duns Scotus, Archimedes, Aristotle, Euclid, Apollonius of Perga, Archytas of Tarentum, Al-Khwarizmi, Al-Kindi, and Gerbert Hispanus.) Suiseth's being on the list means he was known and well thought of 150 years after his work. In fact, he is favorably mentioned through the eighteenth century but is not referred to by nineteenth-century Moritz Cantor, the same historian who wrote off Al-Khwarizmi.

Suiseth, along with the other Scholastics, delighted in

questions involving infinity. "If an object of infinite extent has a finite part which is infinitely dense, would the whole be of infinite density?" This is far removed from the type of infinitesimal analysis that led sixteenth- and seventeenth-century mathematicians to the calculus, but it bears close resemblance to the kind of thing that seems to intrigue modern mathematicians and amateurs.

The work of these Merton scholars was extended at the University of Paris by their pupil, Nicole Oresme, who was also a bishop and adviser to King Charles V of France. Oresme represented quantities geometrically in a rectangular co-ordinate network, which he borrowed from such classical astronomers as Ptolemy. The Chinese also used a rectangular co-ordinate system for both celestial and terrestrial maps.

His system was not really an analytic geometry, because he did not represent algebraic expressions, as such, by geometric curves, and vice versa. That is, he did not reach the stage of saying, as we do now, that the equation $y = x^2$ represents a parabola, or $xy = 4$, a hyperbola, etc. But Oresme's work was certainly close to what we would call co-ordinate or analytic geometry. And his description of relationships between variable qualities anticipated remarkably the modern idea of function—particularly what we call a "linear function"—"a quality uniformly difform is one in which, when any three points (of extension) are taken, that proportion of the distance between the first and second to the distance between the second and third is the same as the proportion of the excess in intension of the first over second to the excess of the second over the third."

By using areas of geometric figures to represent quantity of a physical quality, Oresme also anticipated the basic idea of the integral calculus. He plotted one "quality," velocity, for example, against time, and argued that the area of the figure would represent distance.

Some of Oresme's work with a co-ordinate system had

been covered a few years earlier in Italy by a Franciscan monk, Giovanni di Casili. But the latter's explanation was particularly mystifying in a time when recondite explanations were the order of the day, and Oresme's writings were studied by Italian scholars, in particular, Blasius of Parma.

Blasius was called by a contemporary "the most universal philosopher and mathematician" of the late fourteenth century. He wrote on weights, center of gravity, displacement of fluids—all of which results he developed in the form of Euclidean tradition—that is, propositions were stated and "proved" as theorems. In fact, Blasius even couched his astrological predictions in terms of propositions and corollaries, thus lending them an appearance of a mathematical rigor. As for many mathematicians of his century and later, these astrological forecasts represented, at least, a nice addition to the pay of a teacher, and, usually, a source of reputation.

In June 1386, during the war between the houses of Carrara and Della Scala, Blasius predicted that, if the former engaged in battle at the moment, they would win and take their opponents prisoners. At first, on the contrary, the Scaliger troops routed part of the Carrara army, pursuing it to the town walls. A bystander, and evidently a skeptic, derided Blasius for his prediction. He only replied, "You're crazy. Either it will come out as I said, or the heavens will fall." Sure enough, the unrouted portion of the Carrara army then took the pursuers in the rear. Caught between the two forces, the soldiers of the enemy were taken prisoner, the prediction was fulfilled, and Blasius, presumably, would sneer at those of little faith.

Blasius taught mathematics and moral and natural philosophy at Bologna, Padua, and Parma, and seems to have experienced the usual vicissitudes of teachers of his time. About ten years after his prediction in the war, just mentioned, he came under suspicion of heresy and was suspended from his academic position. He speedily recanted, apparently torture was **not** needed, and was reinstated. At about the same time his popularity as a teacher was declining. On one occasion his lectures were boycotted because he refused to allow Vittorino da Feltre (later a famous educator) to study under him. (Feltre seems not to have had sufficient funds to pay the fees.)

In general, the financial situation of the nonclergy teachers was precarious, at best. It became customary for mathematics professors to charge for the solution of problems and to jealously guard any discoveries he made. He might convey the information to his own student or associates, but they were honorbound not to pass the results along. Thus a statement such as the following was not unusual: "Declaration, that I shall teach no man any of the mathematical sciences, viz. astronomy and geometry, either theoretical or applied, which I learned from my teacher R. Elijah Bashyazi as long as he lives, unless he gives me permission to do it. May God grant him long life, amen."

The breaking of such a pledge led to some lively controversies in the sixteenth century.

The sailor of Chaucer's time was typical of the older breed. He kept in his head all the information he needed to ply his way along the coast. But Chaucer was also aware of the possibilities for more accurate navigation—in particular, of the astrolabe, which he described in detail for his son.

The astrolabe had been known and used on land for many hundreds of years but until the thirteenth century was considered too valuable an instrument to be risked at sea. And anyway, there was no general acknowledgment in the West of its usefulness in navigation. About 1200 some of the Italian city-states—Venice and Genoa, in particular—were expanding their commercial ventures. Merchants and traders recognized the need for improved navigational aids for captains expecting to be out of sight of land for extended periods. Other members of the community were ready and willing to cater to the needs of these

An early cartographer and his apprentice. The presence of the globe suggests that most learned folk knew that the earth was globular in shape long before Columbus made his voyages. The problem: to translate the data from the globe to a flat map, with predictable and regular distortion.

highly respected (and wealthy) citizens. Their efforts brought better instruments, collections of sailing directions for the Mediterranean and Black Seas, reasonably accurate charts for use with the improved compasses, and mathematical techniques. In connection with the last mentioned, the new arithmetic of the algorists began to pay off, for even with the improved instruments and maps, the mariner did need to perform some rapid calculations.

What passed for charts and maps in the West to this time were quite useless for navigation. The "maps" were more representations of religious themes than of portions of the earth's surface. Even when the map maker made an effort to represent faithfully the terrestrial features, he might fit his maps to the size and shape of the parchment at hand. The Moslems had preserved and improved the map-making techniques of the Greeks, but these ideas had been generally neglected in the West.

By the end of the thirteenth century, the situation had begun to improve. Italian maritime charts were scaled with reasonable accuracy and complemented with networks of rhumb lines in thirty-two directions. These charts covered the areas regularly visited by Italian merchants—the Mediterranean and Black Seas areas, the northwestern coast of Africa, the English Channel, Ireland, and Scotland. The pilot needed rulers and dividers (hitherto used only by the geometer, architect, and surveyor) and a bit of arithmetic.

Majorca, where early Catalan colonizers had established a seafaring tradition, became a center of chart making in the fourteenth century. The interest of the Catalans was supplemented by the knowledge and techniques of Jewish astronomers and instrument makers who remained after the Moslems were ousted. This collaboration produced charts and maps for many a maritime venturer, including the kings of Aragon and France.

Raymond Lull was the most famous of these Majorcans—better remembered for his missionary efforts on behalf of the Moslems than for his not insignificant contributions to the development of mathematics and in the field of literature (he wrote a novel describing a Utopia). His manual on mathematical navigation was written in the question and answer form popular in the late thirteenth century. For example, Question

192 went like this: "How do sailors measure their mileage at sea?" Lull's answer: "Sailors consider the four principal winds, namely east, west, south and north, and the other four winds which derive from the first, namely northeast, southeast, southwest and northwest. And they consider the center of the circle at which winds make angles. And, supposing a ship sails on an east wind 100 miles from the center, so many miles does she make on the southeast wind. And, for 200 miles, twice the number by multiplication. And they know how many miles there are from the end point of each 100 miles east to the corresponding points southeast."

He did include a diagram together with some additional explanation, which is needed before Lull's explanation, even with the tables he gave, makes any sense. He uses the term "mile," and probably meant the "little sea-mile" which was about 4,100 English feet. But there were in use also the "Arab" mile and the "Italian" mile, which eventually made for no little confusion when mariners began taking seriously these instructions on navigation.

The Moslems had long sailed the Indian Ocean and had developed techniques for getting across without clinging to the coast. But their innovations were little noticed in the West until the fifteenth century, despite the publicity given them by that best known of travelers, Marco Polo.

Marco Polo first learned the art of navigation at home in Venice, and then went with his father and uncle to the court of the Great Khan. He learned the Mongol language and became a civil servant, trusted even to the point of being put in command of a small fleet on the China Sea. When the Polos were finally ready to return home, they were asked by the khan to escort a Mongol princess who was being sent as a bride to a prince in Persia. Marco argued for a sea voyage as quicker and just as safe (their journey out had taken three years), and off they went, along sea routes used by Moslems since the eighth century. On the voyage home Marco had opportunities to converse with Indian Ocean pilots and navigators, obtaining a copy of their "sailing directions" and knowledge of Moslem navigation methods, including use of the stars.

But real appreciation for these techniques, in fact for most of the navigation developments of the thirteenth century, came only with a great surge of exploration, following the discovery of a sea route to India. Most mariners until then were content to sail reasonably safely from one European port to another, or along the West African coast where there were valuable fisheries. But to sail out into the middle of the Atlantic? Hardly. There was no profit to be gained.

Intellectual, particularly mathematical, activity did not cease outside Europe once Western scholars, princes, merchants *et al* became aware of what was happening in the rest of the world. Egypt experienced a cultural revival under the Mameluke house in the thirteenth century. At least one genuine mathematician, Mizrahi, was active in the old Byzantine realm. There was the mathematical and astronomical activity in Bagdad under the Mongol conquerors, and far to the East, also in the Mongol realm, China produced a couple of creative mathematicians in the thirteenth century.

But these were temporary brightenings of fading cultures. The real initiative had passed to the people of Western Europe, and they would carry the action for some five hundred years until the United States, joined later by the Soviet Union, and then China, began to make significant contributions.

EPILOGUE

Isaac Newton once said that if he had seen farther than others it was because he had stood on the shoulders of giants. Other mathematicians, scientists, and, indeed, artists, writers, and philosophers of his time and later could make the same acknowledgment. Newton might also have noted that he lived in a cultural climate that was considerably more favorable to his inclinations than was that of another era. That is, no creative person operates completely in a vacuum. He is part of his time, and of the continuum of civilization.

These "giants" of the thirteenth century stood on the shoulders of the Gerberts, the Adelards, the Fulberts in the West, the many Moslems and Jews, and the translators. They, in their turn, had benefited from the great work of the Greeks, who had been influenced by the Babylonians and the Egyptians. The particular emphasis in mathematics toward the description and explanation of a variety of natural phenomena reflected the spirit of the times and set the stage for the refinements of later centuries.

The projective geometry of Desargues and Poncelet was based on several centuries of experimentation by painters, map

Epilogue

Copernicus observing an eclipse of the moon during his stay in Rome (1500). From Louis Figurier, Vie des Savants. *It seemed to me an appropriate final illustration for this volume, as we look ahead to the mathematical and scientific innovations of the sixteenth and seventeenth centuries.* Courtesy Scripta Mathematica

makers, and just "dabblers." The calculus of Newton and Leibniz grew out of investigations and approximations by Suiseth, Oresme, and others, and has a tradition going back to Archimedes and Zeno of Elea. Descartes' analytic geometry represented a culmination and refinement on the efforts of Oresme, of Moslem and Chinese map makers, and again back to the Greek geometers. In astronomy Tycho Brahe and his successors used and improved on the tables of the Moslems, and many of Kepler's results came inadvertently as he investigated number properties in the tradition of Boethius and the Pythagoreans. Contributing to the success of all these were the algorists and their "new arithmetic."

The chronicler of the development of mathematics must break somewhere, and this time of intensification of activity in the West seems a good place. I can end on a note of expectation, echoing the words of Professor Thorndike: "And so, with a spirit of inquiry and ingenuity which may be a little naive but is equally sincere, they look forward not backward, and out on the world of phenomena as well as into their books."

INDEX

Abacus (abaci), 55, 68, 71, 72, 80, 86, 106, 108, 109, 114–18
Abbassid al-Saffah, 30
Abelard, Peter, 74
Abraham bar Hiyya (Savasorda), 101
Abu Kamil, 125
Acceleration, uniform, 134–35
Adab, 31
Adelard of Bath, 78–80, 83, 143
Adelbold, 68
Adoration of the Magi, The (Dürer), 89
Afghanistan, 39
Agriculture, Chinese mathematics and, 58
Al (in Arabic names). *See under* second term, *e.g.*, Al-Biruni . . . *see* Biruni, Al-
Alchemy (alchemists), 8, 120, 133
Alcuin of York, 13, 19, 22–24, 86, 99, 109
Alexander III, Pope, 90
Alexander the Great, 3–4, 10
Alexandria, 1, 2–3, 10–11, 26, 27, 29, 46, 85, 86; destruction of library, 10–11, 26, 27, 29, 46, 86; first period, 2–3, 10–11; second period, 3, 11
Algebra, 4, 5, 34, 36–38, 58
Algebra (Al-Khwarizmi), 23, 34, 36, 80
Algorisms, 68, 109, 114–20, 139, 144
Almagest (Ptolemy), 5, 40, 96, 97, 100–1
Alphonso X ("El Sabio"), 101–2, 120
Alphonsine Tables, 102
Anuyogas, 49
Apollonius (of Perga), 2, 11, 55, 135
Aquinas, St. Thomas, 77
"Arab horse problem," 37
Arabs, 4, 26–46, 49, 66, 94, 99, 101 (*see also* Moslems; specific developments, individuals, places); language, 30, 39, 49, 66, 101; and numeration, 80, 99, 104–6, 107, 108
Archimedes, 2, 3, 5, 10, 61–62, 96, 100, 131, 135, 144

Archytas of Tarentum, 96
Arezzo, Italy, 75
Arian Christians, 16–17, 100
Aristotle, 30–31, 43, 71, 76, 77, 80, 97, 100, 112–13, 135
Arithmetic, 6, 14, 22, 50, 58, 74 (*see also* specific developments, individuals, people, places); new (*see* New mathematics); numeration systems, 103–21 (*see also* Numeration systems)
Arithmetica (Diophantus), 4
Arithmetical Classic of the Gnomon and the Circular Paths of Heaven, The, 51
Armenia, 91, 94
Art(s), 71, 122 (*see also* Painters); liberal, 22, 71, 72, 74; Moslems and, 28, 29, 41, 45
Arthur, King, 19
Articuli, 109
Aryabhata, 47, 48
Asceticism, 18
Ash'ari, Al-, 43
Assyrians, 37
Astrolabe, 38, 39, 124, 138
Astrologer, The (Scherr), 9
Astrology, 8, 9, 33, 34–35, 48, 50–51, 79, 97, 99, 120, 123, 133, 137. *See also* Astronomy; Planets
Astronomer, The (Dürer), 79
Astronomy, 2, 5, 8, 9, 12, 19, 32–33, 38–39, 40, 47–49, 58–59, 66, 68, 71, 83, 84, 92, 100, 101, 120, 130, 138, 144. *See also* Astrology; Planets
Athanasius of Balad, 38

Babylonians, 8, 38, 86, 111, 143
Bacon, Roger, 80, 91, 98, 133, 134
Bagdad, 33, 35, 40, 82, 85, 91, 92, 142
"Baptized sultans," 96–97
Base-number numeration systems, 104ff., 111ff.

Index

Batu (Mongolian commander), 90
Bede, "Venerable," 13, 19–21, 66, 71, 86, 109
Benedict, St., 18
Benedict IX, Pope, 130
Benedictine monasteries, 18, 19, 25
Benjamin of Tudela, 83–85, 87
Bhaskara, 50–51, 113
Bible, the, 38, 71
Biruni, Al-, 29, 34, 39–40, 47–49, 50, 54, 82, 83, 86, 107
Blasius of Parma, 122, 137–38
Boethius the Roman, 13–18, 23–24, 71, 80, 144
Bologna, university in, 75, 137
Bolyai, John, 43
Bookkeeping, 74, 86
Brahe, Tycho, 144
Brahmagupta, 47, 48, 113
Buddhism, 52–54, 77
Bulan, Prince, 85
Bureaucracy, 98; Chinese, 58–59
Byzantine Church. *See* Orthodox Eastern Church
Byzantine Empire, 31, 85, 87, 94–95, 109, 142
Byzantium, 31, 85. *See also* Constantinople
Byzas, 93

Caesar, Julius, 10, 11
Cairo, 33, 85
Cajori, Florian, 49–50
Calculating methods and systems, 55–57, 68, 72, 80. *See also* Algorisms; Numeration systems; specific kinds, people
Calculus, 52, 123, 135, 136, 144
Calendars, 48–49, 58, 71
Cantor, Moritz, 36, 43, 135
Cardan, J., 135
Carthage, 95
Cartography (maps, mapping), 5, 9, 34, 39, 59–60, 61, 139–40, 143–44
Cathedral schools, 71–72, 74
Charlemagne, 13–14, 19, 21–22, 23, 29, 65, 82, 95, 99
Charles V, King (France), 136
Charles the Bald, 99
Chartres, 68, 71–72, 73
Chaucer, Geoffrey, 15, 122, 124, 138
Chess problems, 39–40
Chia Tan, 60
China (the Chinese), 26, 31, 34, 39, 51–64, 77, 81, 82, 85, 86, 88, 92, 108, 111, 112, 113, 134, 136, 142, 144
Christ. *See* Jesus (Christ)
Christianity, 11, 16–17, 18–22, 26, 29–30, 66–72, 76, 81, 82, 86–102 (*see also* Church, the); Crusades, 82, 93–94, 98
Chronology of Ancient Nations, 39
Chuquet, Nicolas, 113–14
Church, the, 66–72, 76, 82, 86–102, 118–20, 123, 126–27. *See also* Christianity;
Monasteries; specific denominations
Chu Shih-Chieh, 57, 62, 63, 64
Circles, value of pi and, 61–62
Cleopatra, 10
Clergy (priests), 66–72, 76, 81. *See also* Church, the; Monasteries
Clerks, 66, 72
Columban, St., 19
Columbus, Christopher, 139
Compano da Novara, 134
Compass, use of, 139
Computer techniques, Chinese and, 55–57
Comtine, Rabbi Mordecai, 116–18
Confucianists, 58
Consolation of Philosophy, The, 16
Constantinople, 82, 85, 87, 88, 93–95
Co-ordinate systems, 55, 69; rectangular, 136–37
Copernicus, Nicolaus, 77, 144
Coptics, 29–30, 109
Cordova, 32, 99
Counting rods (counting boards), 55–56, 59, 86. *See also* Abacus (abaci)
Craft guilds, 72–75
Cross multiplication, 116–18
Crusades, 82, 93–94, 98
Cubes, 6, 7
Curves, 55
Cyril, Bishop of Alexandria, 11

Dakiki (poet), 43–44
Dante Alighieri, 122, 123, 130–131; *Divine Comedy*, 130
Dantzig, T., 11
Dark Ages, 13–24, 44, 65ff. *See also* Middle Ages; specific developments
Decimals, use of, 55, 104, 106, 107, 109, 112, 119, 120
Dedekind, Julius, 43
"Deficient" numbers, 23, 24
Desargues, Gérard, 143
Descartes, René, 91, 144
Description of India, 39
Diatonic scale, Pythagorean, 17–18
Digital computers, 55–57, 72
Digital reckoning, 19, 105, 109, 111
Digiti, 109, 118
Diophantine problems, 57
Diophantus, 2, 4–5, 11, 36, 121
Disciplina Clericalis, 83
Dodecahedron, 6–7
Dominicans, 98, 126
Dondi, Giovanni (John de Dondes), 122–23
Dream Pool Essays, 54
Dreams, 30–31
Dumbleton, John, 135
Dungal (Irish monk), 22
"Duodecimal Society," 104
Duplation multiplication method, 115–16, 118
Dürer, Albrecht, 25, 79, 84, 89

Index

Earth, early beliefs and theories of, 48, 76, 79, 129, 133, 139
Easters, determination of, 21
Eclipses, 4, 5, 38, 48
Egypt (Egyptians), 3, 10, 29–30, 86, 107, 108–9, 116, 142, 143
Elements (Euclid), 4, 7, 36, 80, 96
England (the English), 13, 19, 72, 78–80, 83, 133–36
Equations, 33, 49, 56–57
Essenes, 18
Euclid, 1–2, 3–4, 36, 42, 80, 96, 100, 127–29, 135, 137; *Elements,* 4, 7, 36, 80, 96
Experimental science, development of, 77–80, 120, 123, 131ff.

Fa Hsien, 52–53
"False Position, Rule of," 56–57
Feltre, Vittorino da, 138
Feng Lao, 59
Feudal system, 58, 72–74; decline of, 66
Fibonacci. *See* Leonardo of Pisa (Fibonacci)
Figurate numbers, 14–16
Figurier, Louis, 144
Finger (digital) reckoning, 19, 105, 109
FitzGerald, Edward, 42
Fractions, 111, 119
France, 69, 74, 82, 91
Franks, 21–24, 29, 82
Frederick II, 97–99, 123–25, 127, 131–33
Fulbert, 68–69, 71, 72, 143

Galen, 92, 100
Galileo Galilei, 18, 77
Gaon, Rabbi Saadia, 116
Genghis Khan, 91
Genshin, 130
Geography, 5. *See also* Navigation
Geometry, 36, 38, 42, 71, 113, 138; analytic, 123, 136, 144; Euclidean, 4, 5–6, 7, 36, 42, 127–28, 137; Greeks and, 1–2, 14, 19, 22, 58, 144; non-Euclidean, 42–43; projective, 123, 129ff., 143
George, "Bishop of the Arab Tribes," 38
Gerard of Cremona, 36, 100–1
Gerbert of Aurillac (later Pope Sylvester II), 16, 24, 66–70, 71, 78, 80, 86, 109
Ghazzali, Al-, 43, 77
Ghubar numerals, 108, 109, 118
Giotto di Bondone, 122, 123, 130, 131
Giovanni di Casili, 137
"Golden section, the," 7
Grammar, 71, 75
Grand Master of the Assassins, 40
Great Renaissance, 13, 65, 71
Greece (the Greeks), 1–12, 14, 27, 30, 31, 33–35, 37–39, 44–46, 67, 81, 86, 93, 95–97, 100, 130, 143, 144 (*see also* specific developments, individuals); and numeration system, 108–9, 112–13, 121
Greek Orthodox Church. *See* Orthodox Eastern Church

Grosseteste, Robert, 133
Guild Merchant, 72, 74
Guilds, 72–75
Gur Khan, 90

Hadrian, Emperor, 23, 24
Hakam II, 32, 33
Han-Lin Academy, 54, 64
Harmonies of the World (Kepler), 7
Harsha (king), 53
Harun al-Rashid, Caliph, 82
Heavenly bodies. *See* Planets
Helmholtz, Hermann L. F. von, 18
Heresies, 88, 123, 127, 137
Hero (of Alexandria), 131; formula, 5; *Pneumatics* of, 96
Heytesbury, William, 135
Hindu-Arabic numerals, 80, 104–6, 107–9
Hindus, 26, 31, 36, 47–52, 80, 104–6, 107, 108, 109. *See also* India (Indians)
Hipparchus, 2, 5
Hippocrates, 72
Historical Records of the Yuen Dynasty, The, 64
History of the Church of the English People (Bede), 19
History of the Four Sons of the House of Gingiz, 32
Holy Roman Empire, 102; emperors, 97
Horace, 71
Hrotsvitha, 23–24, 25
Hsuan Chuang, 53
Hulagu (Mongol khan), 92
Humanism, 122, 130
Hypatia, 11–12, 29, 86
Hypsicles, 4

Ibn al-Nadim, 30–31
Ibn Ezra, Abraham, 30, 83
Ibn Qutaiba, 31, 33
Ibn Rash (Averrhoes), 98
Ibn Vahab, 85–86
Icosahedron, 6
Idrisi, Al-, 34
Indeterminate equations, 33, 49
India (Indians), 18, 26, 38, 39, 47–51, 52–54, 55, 57, 68, 77, 81, 85, 86, 87, 88, 107, 108, 111, 113, 142. *See also* Hindus
Indian Ocean, 141–42
Indochina, use of zero in, 113
Ine, King (Wessex), 95
Infinity, 136
Inheritance problems, 34, 36–37, 116
Inquisition, 123, 126–27
"Interpolation techniques," Chinese, 57
Ireland, 70; monasteries, 18–19, 20, 22
"Isaac the Jew," 82
Islam, 12, 27, 29ff., 43, 90. *See also* Moslems
Italy, 69–70, 72, 75, 94–95, 98, 108, 127, 137–40 (*see also* specific places); Renaissance, 2, 34, 122, 130

Index

Janus, 19
Japanese, the, 26, 130
Jarrow, university in, 19
Jesus (Christ), 16, 29, 76
Jews, 4, 21, 26, 29–30, 34–36, 38–39, 43, 44, 77, 82–85, 96, 98, 100, 101, 108–9, 116, 120, 140, 143; prejudice against, 82, 85, 87, 120, 127
John of Carpini, 90–91
John of Gorze, 99
John of Palermo, 125
John of Saxony, 101
Jordanus Nemorarius, 113, 126
Julian of Toledo, 99
Justin, Emperor, 16

Kashi, Al-, 112
Kepler, Johannes, 7, 8, 18, 144
Key of Astronomy (Al-Biruni), 48
Keys of the Sciences (Al-Khwarizmi), 31–32
Khazars of Crimea, 85
Khwarizmi, Muhammed ben Ahmad Al-, 31–32
Khwarizmi, Muhammed ben Musa Al-, 23, 32, 34, 35–38, 80, 83, 85, 86, 104, 109, 114–15, 116, 135
Khwarizmshah, 39
Kibla, 33–34
King Roger's Book, 96
Kublai Khan, 64, 92
Kumara (king), 53
Kuyuk (Mongolian khan), 90

Language(s), 30, 37, 46, 49–50, 55, 107
Leibniz, Wilhelm, 52, 135, 144
Leonardo of Pisa (Fibonacci), 36, 99, 101, 104, 113–15, 120, 122, 123–26, 134
Letter writing, 68, 75–76
Liber abaci, 99, 101, 114, 120
Liberal arts, 22, 72, 74
Lilavati, 50–51
Linear functions, 136
Literature, 23, 49–50, 68, 71, 101–2
Liu Hui, 62
Li Yeh, 64
Lobachevski, N. I., 43
Lorraine, 71, 72, 99
Louis IX (St. Louis), 91
Lull, Raymond, 140–41

Magic squares, 35
Magnets, 134
Magnification, mirrors and, 133–34
Mahaviracarya (Mahavira), 47, 49
Mahmud, Sultan, 39
Maimonides, 44
Majorcans, 140–41
Ma'mun, Al-, Caliph, 30–31, 35
Manuel (Byzantine emperor), 88
"Map of both Chinese and Barbarian Peoples . . . ," 60
"Map of the Tracks of Yu the Great," 60

Maps (mapping). *See* Cartography
Mark Antony, 10
Mathematical Collection, The (Ptolemy), 5. *See also Almagest* (Ptolemy)
Matrix notation, 57
Mayan civilization, 86, 112, 113
Mecca, 27, 33, 43
Medicine, study of, 71–72
Medina, 27, 43
Menelaus, 5
Metric system, 104, 120
Metrodorus, 4
Middle Ages, 8, 60, 72ff., 86, 87. *See also* Dark Ages; specific developments
Military, the (warfare), mathematics and, 59, 91
Mirrors, magnification and, 133–34
Mishnat ha-Middot, 38
Missionaries, 19, 87, 88–89, 90
Mizrahi, 118, 142
"Modulus representation," 57
Monasteries (monks), 18–24, 52–54, 65–66, 67, 70, 86, 95, 99
Mongols, 33, 40, 44, 55, 63–64, 82, 90ff., 141
Moors, 100
Moslems, 2, 8, 23, 26–46, 47–52, 57, 65–66, 68, 69, 74, 77–82, 85–88, 90–92, 94, 96–101, 126, 130, 131, 140–43 (*see also* Islam; specific developments); mosques, 28, 41; and navigation, 141–42, 144; and numeration systems, 107, 108–12, 120, 121
Motion, laws and theories of, 7, 8, 48. *See also* Acceleration, uniform; Earth; Planets
Mozarabs, 100
Mudéjars, 100
Muhammed the Prophet (Mohammed, Mahomet), 27, 32, 34, 43
Multiplication, 104ff., 114–18, 121
Music, 2, 5, 14–15, 17–18, 22, 69, 71, 101
Mysticism, 18, 47–48, 49–50, 57–58, 87; Pythagoreans and, 6–7, 8, 16

Nalanda, 53
Natural History (Pliny the Elder), 19
Natural Questions, 79–80
Navigation, 86, 138–42. *See also* Travel
Neckam, Alexander, 134
Neo-Confucianism, 77
Nestorians, 88, 90, 91
New mathematics, 74, 77–78, 102, 104, 114, 118, 120, 123, 127–42, 144. *See also* specific developments, individuals
Newton, Sir Isaac, 52, 97, 135, 143, 144
Nicea, Council of, 21
Nichomachus, 14
Nishirvan the Just, 30
Non-Euclidean geometry, 42–43
Normans, 19, 34, 96–97
Numbers theories, 14ff., 19, 23–24, 110, 121; Greeks and, 6–7, 121

Index

Numeration systems, 55–57, 68, 71, 80, 86, 99, 102, 103–21. *See also* Calculating methods and systems; specific developments, people, systems
Numerical analysis (numerical methods), Chinese and, 56–57

Octahedron, 6–7
Omar Khayyam, 29, 40–43, 44, 62, 121
Omicron, 112
Oresme, Nicole, 122, 123, 136–37, 144
Orthodox Eastern Church (Greek Orthodox Church), 29, 81, 87–89, 93, 94
Ostrogoths, 16–17
Otto I, Emperor (Otto the Great), 23, 25, 65, 66, 99
Otto II, Emperor, 95
Otto III, Emperor, 95
Ovid, 71
Oxford University: Merton College of, 134–36; rise of, 74

Padua, 75, 137; university in, 122
Painters (artists), and new mathematics, 129–31, 132, 143
Palestine, 18, 26
Pappus, 12
Parabolas, 134, 136
Paris, 74; University of, 136
Parma, 137
Pascal, Blaise, 62; "Triangle" of, 5, 62
Paulisa, 49
Paul the Deacon, 22
Pax Mongolica, 91, 93
Pax Romana, 91
"Perfect" numbers, 24, 25
Pergamum, library in, 10, 11
Persia (Persians), 27, 29, 30–31, 39, 82, 88, 95, 109
Perspective, science of, 129–31, 132
Peter Alphonse, 83
Peter of Pisa, 22
Petrarch, 69, 122
Phei Hsui, 60
Pi, value of, 61–62
Pilgrims, 65–66, 82
Place holders, positional number systems and use of, 106–7, 108, 111
Planets (planetary motion and orbits), 7, 8, 18, 34–35, 48, 97. *See also* Astrology; Astronomy
Plato, 87, 96, 121; "Platonic solids," 6
Pliny the Elder, 19
Plutarch, 87
Politics, mathematics and, 8–12, 127, 130
Polo, Marco, 88, 141–42
Polygons, 61–62; regular, 16; star, 16, 18
Poncelet, Jean Victor, 143
Positional number systems, 55, 68, 71, 106, 108, 109ff.
Precious Mirror of the Four Elements, 62, 63, 64
Prester John, 88–90

Problems for the Quickening of the Mind, 22
Ptolemy, Claudius, 2, 5, 60–61, 84, 86, 101, 112, 136
Ptolemy I (Ptolemy Soter), 3–4, 10–11
Ptolemy IV, 11
Ptolemys, 10–11. *See also* specific individuals
"Pure" mathematics, 120–21, 126
Puzzle-problems, 14, 22–24, 37, 39–40, 125–26
Pythagoras, 2, 3, 12, 51, 73
Pythagoreans, 3, 6–7, 8, 17–18, 23–24, 108, 144; and music theory, 17–18; pentagram of, 16; and regular solids, 6–7; theorem of, 5, 7

Quintilian, 19
Quran, the (Koran), 27, 29, 30, 32, 33, 34, 36–37

Radhanites, 82–83
Ramadan, 34
Ratios, 7, 43; Pythagoreans and, 17–18
Ravenna, Italy, 94–95
Razi, Al-, 92
Real numbers, systematizing of, 42, 43
Regiomontanus, 40
Religion, 29, 33–34, 43, 44–45, 46, 47–48, 49, 86–102. *See also* Church, the; specific denominations, developments
Renaissance(s), 65–80, 95, 122–42; "Carolingian," 22–24; "Great," 13, 65, 71; Italian, 2, 34, 122, 130; "Renaissance man," 54, 102, 122
Rheims, 66, 71
Right triangles, 6, 50, 51, 59
Rithmomachia, 110
Robert of Chester, 36, 80
Robert the Pious, King, 69
Roger II, 34, 95, 96–97
Roman Catholic Church, 65, 88ff., 96, 100. *See also* Church, the
Roman era, 3, 4, 10, 11, 14–19, 21–22, 24, 39, 67, 72, 95, 96 (*see also* Roman numeration); collapse of empire, 14, 26
Roman numeration, 68, 71, 83, 102, 104–6, 108, 109, 118, 120
Rabaiyat, The, 42, 43

Salerno, medical school in, 98, 99
Salimbene (Franciscan monk), 127
Samarkand, 32–33
Sapientia (Hrotsvitha), 25
Sarton, George, 30
Scale ratios, Pythagoreans and, 17–18
Scholasticism, 76–80, 87, 123, 133, 134–36
Science, 20, 31–33ff., 43, 47ff., 58, 66, 68, 69, 81ff. (*see also* specific aspects, developments, individuals, people, places); new mathematics and, 129–42; renaissance era, 120, 123, 127–42; and

scientific method (experimental science), 77–80, 120, 123, 131–42
Scot, Michael, 98–99
Sea Island Arithmetic, The, 59
Secretaries, 31
Serapis, 11
Severus Sebokht, Bishop, 38–39, 104
Sexigesimal (base-sixty) notation, 111–12
Shen Kua, 54–55, 58
Shu Hsi, 59
Sicily, 34, 44, 82, 85, 93, 94–96
Silk trade, 86
Sindbad the Sailor, 85–86
Solids, regular, 6–7
Spain, 26, 29, 32, 44, 66, 70, 83, 98, 99–101, 108–9, 116, 127
Spherical trigonometry, 34, 48
Stabius, Johann, 84
Star polygons, 16, 18
Stevin, Simon, 43, 120
Student guilds, 74–76
Suiseth, Richard, 122, 123, 135–36, 144
Sulaiman (Soleyman) the Merchant, 85–86
Sung dynasty, 54–56, 57, 63–64
Symbols, numeration systems and, 55, 103ff., 106–8, 111, 112–21
Synesius, 11
Syracuse, 3, 96
Syria, 29–30, 38–39, 85, 88, 93, 109

Taoists, 57–58
Tartars, 44
Taxes, 31, 58, 59, 72
Tesellations, 45
Tetrahedron, 6–7
Thales, 2
Theodore, Master, 98, 99
Theodoric (Ostrogoth ruler of Rome), 16
Theon of Alexandria, 4, 11
Time computing systems (chronologies), 21, 39, 42. *See also* Calendars
Towns, growth of, 66, 72–74
Trade (commerce), 66, 72, 82–85, 88, 93, 94, 95, 138–40

Translations (translators), 5, 18, 27, 30–31, 34–36, 38, 44, 49, 80, 82, 83, 96–97, 98, 100–1, 109–11, 143
Travel, 54, 66, 70, 78–79, 81, 82–86, 138
Triangles, 62–63, 68; cosmic, Pythagoreans and, 6; right, 6, 50, 51, 59
Trigonometry, 40, 48; spherical, 34, 48
Tsu Chung-Chihl, 62
Turkey, 87, 93
Tusi, Nasiruddim, 40
Tycho Brahe, 144

Uccello, Paolo, 122, 123, 131
Uleigh Beg, 32–33
Ung Khan, 90
Uniform acceleration, 134–35
University centers, 52, 53, 71–76, 122, 126, 134ff. *See also* specific places

Vasari, Giorgio, 131
Velocity experiments, 134–35
Venice, 94, 138
Vernal equinox, 21
Vie des Savants, 144
Vincent of Beauvois, 134
Virgil, 71, 130
Voltaire, 3
Vries, Jan Vredeman de, 128, 132

"Wandering scholars," 70–71, 74
Wenceslaus, King of Bohemia, 90
William of Rubrouck, 91
Witelo, 134
Women: influence in Constantinople, 94; as mathematicians, 59
World maps, 34. *See also* Cartography

York, monastery in, 19
Yuh Hing, 125

Zeno of Elea, 113, 144
Zero, use of, 55, 103, 106–7, 109–11, 112
Zoroastrian religion, 29, 44

CHARLES F. LINN is a graduate of Colgate University and later went to Wesleyan University in Connecticut, where he earned an M.A. and, along the way, gained, as he says, "my first insights into what math is all about, and the idea that everyone can be creative in math at his own level."

Mr. Linn then not only taught mathematics in public schools, but later was the mathematics editor and writer for two nationally circulated classroom scientific newspapers. He is currently teaching at Oswego State College in upper New York State.